SpringerBriefs in Law

More information about this series at http://www.springer.com/series/10164

Tafsir Johansson · Patrick Donner

The Shipping Industry, Ocean Governance and Environmental Law in the Paradigm Shift

In Search of a Pragmatic Balance for the Arctic

Springer

Tafsir Johansson
World Maritime University
Malmö
Sweden

Patrick Donner
World Maritime University
Malmö
Sweden

ISSN 2192-855X ISSN 2192-8568 (electronic)
SpringerBriefs in Law
ISBN 978-3-319-12540-4 ISBN 978-3-319-12541-1 (eBook)
DOI 10.1007/978-3-319-12541-1

Library of Congress Control Number: 2014953257

Springer Cham Heidelberg New York Dordrecht London

Printed on acid-free paper

Springer International Publishing AG Switzerland is part of Springer Science+Business Media (www.springer.com)

Foreword

In contemporary Arctic, there are numerous specific issues, i.e. theories on how to delineate jurisdictional boundaries, regulatory codes to deal with ameliorated commercial shipping, regulations on how to prepare for the possibility of oil spills under Arctic conditions, rules to impose on tour operators and so forth. In short, it is apparent that there is no scarcity of legal framework for the Arctic zone; however, those legal frameworks are likely to give rise to vigorous debates that result in decisions that may well have significant impacts on human activities in the circumpolar north. Underlying such debates, however, are in-depth or broader issues circulating the legal frameworks the Arctic members have employed to operate in specific environmental situations. What has been left out is the inescapable truth that the doors of the Arctic are opening up and any door that is open in the ocean is bound to be utilized by the shipping industry for commercial advantage. The Arctic anticipates an era of 'high politics' marked by aggressive assertion of jurisdictional claims, increasing competition for control and rights over the Arctic's natural resources, a remilitarization of the region, and more or less frequent clashes among leading states active in the Arctic.[1] The Arctic Council has been the conglomeration for the Arctic States, which prescribes 'soft law' to address environmental issues, but it has remained silent on the most pressing challenges facing the region.

[1] Borgerson, Scott G., *Arctic Meltdown: The Economic and Security Implications of Global*, 2008, Foreign Affairs, Published by the Council on Foreign Relations, available at; http://www.foreignaffairs.com/articles/63222/scott-g-borgerson/arctic-meltdown (date accessed 6 August 2014), where the author has added the fact that the situation is dangerous because there are currently no overarching political or legal structures that can supplement for the orderly development of the region or mediate political disagreements over Arctic resources or sea lanes. Then again, '[t]he Arctic has always been frozen; as ice turns to water, it is not clear which rules should apply. The rapid melt is also rekindling numerous interstate rivalries and attracting energy-hungry newcomers, such as China, to the region. The Arctic powers are fast approaching diplomatic gridlock, and that could eventually lead to the sort of armed brinkmanship that plagues other territories, such as the desolate but resource-rich Spratly Islands, where multiple states claim sovereignty but no clear picture of ownership exists'.

Then again, many scholars and observers have relied on the umbrella Convention for answers. UNCLOS, in this context, might be the essential instrument to bring about an order, but considering the unique geographical scenario, it cannot be seamlessly applied to the Arctic. Hence, the Arctic still remains vulnerable and will increase in the level of susceptibility as the ice melts leaving the balance between environmental protection and increased commercialization in an unresolved position.

The book aims at supplementing a critique of the existing international instruments, regional responses and national legislation of the Arctic related to marine environmental protection. With this aim in view, the book satisfies its objective by proposing a Bipartite 'Arctic Council' acting as a conglomeration of the Arctic States and the Flag States with a hybrid 'Arctic treaty', which is an interplay of international and regional response. The 'Arctic Council', apparently, has been highlighted and envisioned as a platform that can provide a significant solution, if modified accurately, to balance sustainable development (marine environment) and international navigation (trade and commerce) in the event of rapid climate change.

In an endeavour to examine the pertinent environmental legal regime of the Arctic, it seemed important to delve into the maritime boundary delimitation issues that involve three major Arctic States. These issues, which subsist in two significant Arctic sea routes, have a subtle connection with the subject of marine protection, which is revealed after a detailed analysis of the geographical issues. While the landscapes are shaping up as a result of global warming, certain Arctic States have risen to the occasion to extend their maritime boundaries in the offshore areas. They have not only resorted to contradicting theories to establish sovereign claims, but also adopted extreme standards and implemented them in national legislation. 'Conflict of law', which in turn distorts the international legal regime, is evident from the comparative study among the national legislation of significant Arctic States. More significantly, this distortion leaves a question on the face of Arctic marine protection. Investigations lead to the fact that the boundary issues have distracted the Arctic States from promulgating a parallel system to safeguard the pristine environment and have left the entire Arctic environmental protection regime in disarray. Inevitable as it is, climate change will accelerate international navigation and break any resistance which operates against 'due regard to navigation' as embedded in the *lex specialis* regime of UNCLOS. Moreover, areas beyond national jurisdiction have not received proper attention and till date none of the zones have been designated as MPAs. On the other hand, the international community, which supports 'freedom of navigation', only seeks commercial advantages of a shorter sea route. There is a vacuum of global concern. Moreover, the international instruments and regulatory conventions portray a lack of respect for the Arctic, which is seen as the 'last ecosystem on earth'. Apart from dealing with inconsistent geographical claims, the Arctic States have responded via Arctic Council, which is an intergovernmental forum established for the purposes of addressing questions of sustainable development as well as environmental issues. With no specific mandates, the five working groups under the Council suffer from low funding.

The 'soft law' character of the 'Arctic Council' has been viewed as a major drawback, and the Arctic legal regime is found to be much less comprehensive when compared to the treaty-based regime that regulates the Antarctic, a region with a very similar environment. As such, discussions have proceeded as to whether the Arctic is in need of a new legal regime, and whether the Antarctic treaty should be a model. What is truly needed is structure, and regardless of which shape the future Arctic legal regime takes, the most important aspect is that the existing 'Arctic Council' must take into consideration the geographical and environmental impacts of climate change and supplement a comprehensive legal order. It is not the single concern of the 'Arctic Council' to consider and supplement a legal order, but it should be a global consideration to work hand in hand with the Council to implement this order. If the shipping industry is to provide support, the 'Arctic Council' will need to provide further clarification concerning many questions, among of which one is, how this comprehensive legal order will correctly balance environmental governance and international trade.

The concept of 'Climate change', by now, has become a cliché when describing the catalyst behind geographical and environmental changes in the Arctic Ocean. As is understood, the impact of this catalyst in the Arctic ocean will, over the next decades, ascertain its transformation from a permanently ice-covered and virtually untraversable area into a seasonal navigable sea.[2] Before investigating the complex commercial implications of the Arctic region, it is important to analyse the controversial geographical issues among the Arctic States. Overlapping claims have provoked some States to put into place domestic laws which contradict international law. The Arctic States relate to theories which proceed in favour of territorial sovereignty as regards to disputed offshore regions, criticized in substance by the international community. This is, to a great extent, distorting the international regime and leaving the shipping industry frustrated. These innovative theories defy the international regime and the prospects of international trade and commerce. A number of MOUs exist which the Arctic States denounce spontaneously on certain events to establish opposing claims in those disputed regions. This complex situation defeats the very purpose of diplomatic relationship and, hence, instigates the Arctic States to act reluctantly towards the acceptance of innocent passage endorsed by international instruments.[3] On the other hand, the Arctic States have committed themselves to the

[2] P.A. Berkman and O.R. Young, 'Science and Government: Governance and Environmental Change in the Arctic Ocean', *Science*, 324, 17 April 2009, pp. 339–340, *Note* that, as the northern ice cap changes in response to our changing climate, the northern passage will increasingly experience conditions considered conductive to shipping and other maritime activities. The melting ice is thawing its way to offshore landscapes which provokes Arctic States to claim extended sovereign control, on the other hand, the international instruments or regulatory conventions do not give proper attention to this part of the globe as the melting ice is inviting international vessel-source pollution.

[3] Janet Pawlak, Gunnar Kullenberg and Chua Thia-Eng, 'Securing the Oceans: Executive Summary' in *Securing the Oceans: Essays on Ocean Governance-Global and Regional Perspectives*, Chua Thia-Eng, Gunnar Kullenberg and Danilo Bonga (eds.), January 2008, Published by GEF, UNDP and IMO in association with the Nippon Foundation, p. 4.

regional soft law 'Arctic Council' approach, the participation of which is voluntary. Then again, international instruments corresponding to Arctic environmental protection against intentional or voluntary vessel-pollution are either based on voluntary approach or relate to only a part of the Arctic. Hence, the Arctic has been left disregarded by the international policy makers. Although the umbrella convention, i.e. UNCLOS has a *lex specialis* provision for the Arctic, it has left a question mark on the face of this convention and other international regulatory regimes as to what extent it can relate to the future Arctic navigable routes which is predicted to be the result of this 'climate change' phenomenon.

The notion of climate change is, in fact, influencing geographical and environmental transformation. The push and pull factors of maritime boundary issues are on one side inhibiting international trade and commerce and on the other side, leaving the Arctic marine environment vulnerable. The increase of both intra- and trans-Arctic shipping, specifically poses great pressures and risks in terms of impacts to the Arctic marine environment, its living resources and its biodiversity, leaving the sea route susceptible. Climate change is not only bending the existing international regime, but also leaving the entire Arctic legal regime in obscurity and an undetermined position. Some scholars prefer stringent policies modelled after regimes which do not require balancing of interest and others support the existing voluntary approach which has not yet reached any success in dealing with safeguarding the sensitive Arctic marine ecosystem. The nexus between geographical issues and the environmental issues needs to be analysed in order to comprehend the changes and gaps in the Arctic legal regime. In short, the existing ocean governance system of the Arctic, i.e. international, regional and national legal regimes needs to be revised and examined to set aside stringency and complication and pave the way for international navigation. The existing legal regime needs to be replaced by an enforceable ocean governance strategy to combat the inevitable changes.

The book has been divided into three parts with eight main chapters comprising the main body. The main body starts with a detailed analysis of the existing Arctic international regime. Since the *lex specialis* provision of UNCLOS is significantly connected to the Arctic, the development and pragmatic applicability has been given detailed focus. The Flag State and Coastal State jurisdictions have been examined in a cursory manner since they constitute the general provisions of international marine environmental law and have been placed before the *lex specialis* analysis to maintain the numerical order of the UNCLOS provisions. The IMO regulatory regime follows this analysis with a study of operational discharges under MARPOL 73/78 coupled with a brief overview of the IMO Polar Shipping Guidelines related to the Arctic. In this chapter, the author has made an effort to extract and incorporate the international regime of deliberate dumping and accidental pollution to embody all categories of existing marine pollution provisions that can be related to the Arctic. Then again, since the Arctic has been portrayed as a pristine and sensitive area, the rapid response regime of 'intervention' for the Arctic has been brought under scrutiny. The philosophy behind the raison d'être of the intervention policy lies in the fact that pristine waters should be kept unharmed and there is a growing need for an organized and parallel interventional policy for the entire Arctic. Following the contemporary international regime of the Arctic, a chapter on the pertinent national legislation embodies the first part. Chapter 3 is restricted to the historical development and a critical analysis of existing domestic legislation of the Arctic States, i.e. Canada, the Russian Federation and the USA, since their conflicts constitute a major part of geographical issues in the Arctic. Other Arctic legislation have been given a cursory analysis. It is also important to comprehend these Arctic national legislation which regulate the NWP and NSR, prior to examining the changes that are taking place in those sea routes as a result of climate change.

The second part of the book is designed to comprise two chapters dedicated to the impact of climate change where the former relates to the geographical issues and the latter pertains to the environmental issues. The author is of the view that in order to understand the modifications on the Arctic due to the results of 'climate change' followed by the distorting influence it has on the international regime; it is significant to research into the different theories of geographical issues and to delve into regional responses of the environmental aspect. The inherent reason for highlighting these two issues lies in the relative interconnections that they comprise which is essential to comprehend before discussing the recommendations for a completely new regime in the Arctic. The findings of this part also constitute a segment of the analytical part in the conclusion. Finally, the first half of the final part emphasizes on the 'what', 'why', 'how' and 'which' questions in dealing with Arctic ocean governance.[4] This chapter investigates significant reasons underlying the interest of international trade in the Arctic and the rationale behind balancing commercial interests with interests to safeguard the marine environment. Prior to proceeding to the analysis with striking of the so-called 'balance', a chapter is dedicated to the understanding of the core concept of the 'stakeholder theory'. CSR is intrinsically connected to ocean governance and 'stakeholder theory' is an authoritative concept in relation to the Arctic. This leads to the final chapter where alternatives are examined which can embody an interplay among international, regional and domestic facets and act as a new legal regime which can restrain further distortions due to climate change and at the same time protect the pristine environment of the Arctic.

This book does not include any quantitative statistics of different types of marine pollution followed by the effects of global warming based on scientific analysis, sustainable development related to the indigenous peoples of the Arctic or safety aspects of navigation; such an undertaking would venture beyond the scope of this work. The book, moreover, does not relate to land-based pollution of the Arctic and is limited to the discussion of vessel-source pollution. Although there is

---

[4] The four questions are: what are the economic incentives in the Arctic? Why is there a need to strike a balance between commercial implications and environmental protection? How can this goal be achieved and is there any regime that is closely connected? Which approach is more pragmatic and can be undertaken to strike the desired balance?

an effort to cover international instruments on operational discharges, accidental pollution and deliberate dumping, the analysis centres around the significant instruments which are internationally commendable, directly or technically applicable to the Arctic 'ice-covered' areas, have been ratified by the Arctic States or prescribe bilateral or multilateral cooperation which has the possibility of being implemented via existing 'soft law' approaches in the Arctic. With regard to the chapter on domestic legislation, the examples given are restricted only to three major Arctic States at an in-depth national level and consist of a detailed analysis with a comparative study among them.[5] Those in the authors' opinion are sufficient to understand the 'conflict of laws' and the inconsistencies that persist in international law. The analysis of the national legal framework for marine environmental protection has been limited to the most coherent yet existing instruments of those three States. This is important to understand the existing legal regime of the NWP and NSR (comprising the Arctic sea route), which are the focal points of the book. Hence, in analysing the geographical issues due to climate change, the author elaborately highlights the overlapping issues that exist in those two routes and the domestic theories propounded therein. Then again, there is a plethora of different organizations and institutions that are currently active in the Arctic region, which will not be examined, rather mentioned when necessary. In this regard, the only focus is on the Arctic Council, although brief attention is given to the Arctic regional instrument OSPAR related to marine dumping, since it covers a part of the Arctic. In examining the economic incentives, the thesis does not provide a detailed list of benefits, rather adheres to the advantages of NWP and NSR as shorter sea routes, since it is an actual result of climate change.[6]

The book comprises a qualitative research method for comprehending and scrutinizing various perspectives and issues relating to climate change and the Arctic legal regime. This is followed by a descriptive study and a critical analysis of legal instruments. In brief, the authors have resorted to the dogmatic legal method pertaining to the legal analysis of available resources. The primary sources for the research analysis are coherent provisions of maritime law, both within the areas of domestic and international jurisdiction including but not limited to relevant international conventions, regulatory regimes, applicable treaties,[7] interpretation of judicial decisions and the related jurisprudence. Secondary sources consist of book

[5] This is done to maintain consistency in the chapters to come and so that the analysis remains focused on the NWP and NSR, the commercial implications of which is later balanced with the environmental aspect.

[6] The perspective has been limited to climate change as the central element of discussion and its impact on the existing legal regimes.

[7] In order to achieve the main objective of the thesis, the author has focused on instruments and institutions that are active at a multilateral level.

reviews,[8] chapters in books, journals, compilation of articles, digests, official websites of international organizations, magazines, reports and newsletters.

The aim of this book is to propose a new and unique legal regime for the Arctic under the contemporary regional organization, which is to a certain extent modified, taking into account both the coexisting international and regional perspectives and the concept of good ocean governance as a part of CSR. However, emergence of a new legal system would have an adverse effect on other systems, making it important to set a system of balance, so that it could be in line with the emerging international trade and commerce. To propose a new set of rules, it was primarily essential to examine a series of *status quo* legal layers exclusively related to the Arctic or the ice conditions that subsist today.[9] Although the tradition of commencing with an analysis of pertinent international law has been maintained, it was, eventually, important to observe how these provisions were implemented in the national layer.[10] Since it is the Arctic that is in issue, it was impossible to proceed without revising and scrutinizing the Canadian legislation, i.e. AWPPA which is the only legislation of its kind in the world. Understanding the fact that Canada is faced with opposing interests regarding jurisdictional claims, it was mandatory to focus on other national legislation and their development. An integral part of the main objective of this thesis is to review the impact of climate change, which is, in reality the catalyst behind those 'opposing geographical issues'.[11] In an endeavour to extract and understand the geographical issues and scholarly theories, this book reveals how the environmental issues are left undone at the hands of a Council, which has no specific mandate and will not be able to respond to the increasing ship traffic no matter how stringent the Arctic states are at inhibiting international navigation.[12] All of this leads to the need to understand the aspect of CSR in terms of Arctic Ocean Governance. The 'stakeholder theory' is needed to understand the entities to whom the shipping industry needs to be accountable to. It is apparent that the theory is questionable in itself, but complexity begins to rise when the

[8] As the research topic is current and numerous related topics are still being reviewed and assessed, the research and collection of data is focused mainly on recent technical papers and specialized conferences.

[9] (Question) What is the existing international regime and regulatory regime of the arctic in respect of vessel-source pollution?

[10] (Question 1) What are the pertinent national legislation of the Arctic zone? (Question 2) To what extent of the Arctic area are these legislative jurisdictions applicable?

[11] For the purpose of in-depth analysis, the Arctic sea route has been separately discussed in terms of NWP and NSR, (Question 1) What is the legal status of International Straits in the Arctic? (Question 2) Can NWP be termed as an International Strait? (Question 3) To what extent can Canada claim sovereignty in the NWP based on the Sector theory? (Question 4) How is this distorting the legal regime? (Question 5) How does MOU compromise equality in the NSR?

[12] (Question 1) How does the regional response of the Arctic Council based on a 'soft law' approach contribute as a catalyst of change in the Arctic legal regime? (Question 2) Will the Arctic Council be able to supplement a sufficient yet effective environmental protection management system in terms of the increasing ship traffic?

theory is applied to stakeholders of the Arctic. Then again, to follow with the development of this book, the CSR aspect only relates to protection of the marine environment of the Arctic and does not take into account the human safety and security aspect of shipping or the human element. This is because in trying to fulfil the objective of the book, it was inevitable to create and comprehend the clear nexus between 'geography and environment' of the Arctic. Moreover, the environmental aspect is apparently connected to economic incentives,[13] the element of which needs to be exposed with a view to be balanced with a new legal regime under the domain of a slightly modified organization with a global participation.[14] Finally, the new legal regime (which weighs the balance between interest groups) is explored and assessed and is followed by concluding remarks.

[13] (Question 1) Why is there a need to address a strict Arctic ocean governance? (Question 2) Will this ocean governance be properly addressed if the Arctic is modelled after the Antarctic treaty?

[14] (Question 1) What are the different approaches through which the Arctic legal regime can be addressed? (Question 2) How can the existing approaches be modified and amalgamated into a unique system under the regional response?

Acknowledgments

The authors sincerely acknowledge the work of…

the International Maritime Organization, the World Maritime University, the Arctic Council and the scholars, authors, and researchers of the protection of the Arctic environment.

Tafsir Johansson
Patrick Donner

Contents

Acronyms

ACIA	Arctic Climate Impact Assessment
AEPMT	Act of Environmental Protection in Maritime Transport (Finland)
AEPS	Arctic Environmental Protection Strategy
AFS	International Convention for the Safety of Life at Sea, 1974 (SopS 11/1981)
AMAP	Arctic Monitoring and Assessment Programme
ASPPR	Arctic Shipping Pollution Prevention Regulations
ATCP	Antarctic Treaty Consultative Party
ATS	Antarctic Treaty System
AWPPA	Arctic Waters Pollution Prevention Act, 1970
AWRM	Act of Water Resources Management (Finland)
AWS	Act if Water Services (Finland)
CAFF	Conservation of Arctic Flora and Fauna
CAP	Collaborative Accountability Paradox
CCAMLR	Convention for the Conservation of Antarctic Marine Living Resources
CDEM	Construction, Design, Equipment and Manning
CLC	Civil Liability for Oil Pollution Damage Convention of 1969
CSA	Canada Shipping Act, 2001
CSR	Corporate Social Responsibility
DMA	Danish Maritime Authority
DSDHAE	Decree on Substances Dangerous and Harmful to the Aquatic Environment (Finland)
EEA	European Economic Area
EEZ	Exclusive Economic Zone
EIA	Environmental Impact Assessment
EPA	Environmental Protection Act (Finland)
EPD	Environmental Protection Decree (Finland)
EPPR	Emergency, Prevention, Preparedness and Response
EU	European Union

FSAR	Finland's Strategy for the Arctic Regions
FUND	International Convention on the Establishment of an International Fund for Compensation for Oil Pollution Damage of 1971
GAIRAS	Generally Accepted International Rules and Standards
GATT	General Agreement on Tariffs and Trade
GCTS	Geneva Convention on Territorial Sea and Contiguous Zone, 1958
GDWRM	Government Decree on Water Resources Management (Finland)
GDWRMR	Government Decree on Water Resources Management Regions
GEF	Global Environment Facility
GHREO	Greenland Home Rule Executive Order, 2007
HNS	High North Strategy (Norway)
IAP	Iceland's Arctic Policy
ICES	International Council for the Exploration of the Sea
ICJ	International Court of Justice
IHSCO	International Convention Relating to Intervention on the High Seas in Cases if Oil Pollution Casualties
IMO	International Maritime Organization
IMP	Integrated Management Plan
INS	Integrated Northern Strategy
IO	International Organization
IUCN	International Union for Conservation of Nature and Nature protection
IWCO	Independent World Commission on the Oceans
LC72	Convention on the Prevention of Marine Pollution by Dumping of Wastes and Other Matter, 1972
LOMA	Large Ocean Management Area
MARPOL 73/78	International Convention for the Prevention of Pollution from Ships, 1973 as Modified by the Protocol of 1978
MOU	Memorandum of Understanding
MPA	Marine Protected Area
MPEP	Madrid Treaty on Environmental Protection
NCA	Nature Conservation Act (Iceland)
NCA	Norwegian Coastal Administration
nm	Nautical Miles
NSR	Northern Sea Route
NWP	North-West Passage
OA	1997 Oceans Act (Canada)
OECD	Organization for Economic Co-operation and Development
OPRC 1990	International Convention on Oil Pollution Preparedness, Response and Co-operation, 1990

OPRC-HNS Protocol	Protocol on the Preparedness, Response and Co-operation on Pollution incidents by Hazardous and Noxious Substances, 2000
OSI	Oil Spill Intervention
OSPAR	Convention for the Protection of the Marine Environment of the North-East Environment of the North-East Atlantic, 1992
PAME	Protection of Marine Environment
PES	Protection of the Environment in Svalbard
PSSA	Particularly Sensitive Sea Areas
SDWG	Sustainable Development Working Group
SPA	Sea Protection Act (Finland)
SSA	Societal Sustainability Assessment
SSA	Special Status Area
SSAR	Sweden's Strategy for the Arctic Region
TBT	Regulation (EC) No 782/2003 of the European Parliament and of the Council on the prohibition of organotin compounds on ships
U.S.	United States of America
UN	United Nations
UNCED	United Nations Conference on Environment and Development
UNCLOS	United Nations Convention on the Law of the Sea, 1982
UNDP	United Nations Development Programme
UNEP	United Nations Environmental Programme
United Kingdom	Government of the United Kingdom of Great Britain and Northern Ireland
USGS	United States Geological Survey
WCED	World Commission on Environment and Development
WTO	World Trade Organization
WWF	World Wide Fund

Authors' Biography

Patrick Donner Associate Professor Donner qualified for the bar in Finland (vicehäradshövding) in 1977 and served on the bench as deputy judge and acting chief judge of the City Court of Mariehamn for several years in the late 1970s. Professor Donner served the Sally Shipping group for 12 years, advancing from chief legal counsel and company secretary to deputy managing director with responsibility for the legal affairs of the group, which had ferry operations in Scandinavia and the United Kingdom and cruise operations in the Caribbean. After the Sally Group, he was managing director for Delfin Cruises Ltd. in Finland, operating cruises in the Baltic, after which he ran his own law and management consultancy firm for a few years. During all this time, he taught maritime law (part-time) at the Maritime Academy of Åland and also held numerous non-executive positions on boards of directors of shipping and insurance companies as well as elected public office at local level.

Since January 1995 Prof. Donner has been at the World Maritime University in Malmö, Sweden, first as Associate Professor in Shipping Management and currently as Associate Academic Dean. He is also Head of the Maritime Law and Policy specialisation and during the second half of 2014, he is Acting Vice-President (Academic) of World Maritime University. He teaches international postgraduate students in all commercial aspects of shipping with an emphasis on maritime and commercial law. Professor Donner is an Associate Fellow of the Nautical Institute.

Tafsir Johansson started his career as a civil lawyer of the Commonwealth and joined the World Maritime University (International Maritime Organization of the United Nations) as a research intern in 2013, working with the Environmental Research Group (MER) in its work on various projects as well as the preparatory works for the Arctic Shipping Conference, i.e. 'Shiparc 2014'. He simultaneously worked as a Research Assistant for the 'CSR Maritime' project that pertains to Corporate Social Responsibility and Maritime Governance. Mr. Johansson was employed as a full-time Research Assistant at World Maritime University upon the successful completion of his internship. Mr. Johansson has been hired as a Maritime

Lawyer for MSIIPA (Maritime Spill Incident Intervention Policies and Approaches) under Transport Canada (Government of Canada), Project Officer for PGMS (Good Governance of Multicultural Safety) under International Association of Maritime Universities (IAMU) and Project Officer for SAIL (Sustainable Approaches and Innovative Liaisons) under the North Sea Region programme. He is also a part-time lecturer in the area of Maritime Law and Policy (MLP) at the World Maritime University. Mr. Johansson has co-authored several articles in peered reviewed Journals in diverse areas such as oil pollution in the North Sea, Shipwrecks and Corporate Social Responsibility in the maritime field. As of June 2014, Mr. Johansson commenced his Doctoral Studies in Maritime Affairs at the World Maritime University. Mr. Johansson obtained an all-distinction result in LL.M from the University of Lund in Sweden.

Chapter 1
Status Quo of Arctic International Instruments

International environmental law has been augmented during the past few decades, and in consequence there has been an emergence of several quintessence principles that in turn provide for a framework of customary environmental law. These core principles, though the list of responsibilities is not exhaustive, arising from state practice, have been incorporated in international environmental instruments. They comprise the underlying framework for global marine environmental protection which extends to the Arctic as well.[1] International law in the Arctic has been inspired by areas beyond national jurisdiction and by Coastal States' rights in the offshore regions and other rights regarding resources and navigation. Hence, the forms in which Arctic states administer their territories have been increasingly influenced by international law principles relating to prescriptive coupled with enforcement jurisdiction and environmental protection.

[1] Donald R. Rothwell and Christopher C. Joyner, "Global environmental protection instruments and the polar marine environment" in *Protecting the Polar Marine Environment*, Davor Vidas (ed.), 2000, Cambridge University Press, pp. 57–77. Note that, a major part of international environmental law is aimed at regulating environmental problems by setting common international standards and objectives for prevention or mitigation of harm. It also strives to provide a flexible rule-making process that can permit flexible and regular amendments, since technological and scientific developments require such an approach. However, the provisions on the protection of the marine environment as enshrined in UNCLOS are vitally important. They express principles of international environmental law and provide a framework for establishing a broad and clear structure for the law. UNCLOS is viewed by some as the most significant and comprehensive international environmental agreement that exists today.

T. Johansson and P. Donner, *The Shipping Industry, Ocean Governance and Environmental Law in the Paradigm Shift*, SpringerBriefs in Law,
DOI 10.1007/978-3-319-12541-1_1

1.1 UNCLOS Provisions

Although the suitability of the provisions incorporated in UNCLOS has been questioned in the past[2] and the Arctic itself has been delineated by the media as an area void of international regulations,[3] under the contemporary legal pretext it is by now acknowledged that the Arctic Ocean and the maritime activities do fall within the wider dimension of UNCLOS. Indications have been made to the fact that the UNCLOS was to structure a "Charter of the Ocean" that could act as an elementary framework convention giving insight into major issues of the entire ocean space.[4] With an objective to understanding the international regime of the Arctic, it is important to venture into the prescriptive and enforcement jurisdictions of the Flag State, Coastal State and Port State. More recently, the buzz on the Arctic is climate change and the doorway for international navigation it reveals, so to speak. This legal jurisdictions aforementioned when analysed and extracted, lucidly communicates the basic understanding of the legal regime of international navigation in the Arctic. Hence, Part XII is the part and parcel of the "Charter of the Ocean" with Article 234 as the heart of this integral part as regards the Arctic.

1.1.1 Reference to Part XII of UNCLOS on General Obligations

Articles 192 and 194 as embodied in Part XII of UNCLOS provide a general obligation for every State to protect and preserve the marine environment (in accordance with a State's capabilities) from intentional vessel-source pollution balanced with the reaffirmation of the right of States to exploit their natural resources subject to adopting adequate measures to prevent, reduce and control such pollution.[5] Reference has been drawn to "fragile eco-system", a division to

[2] *See* Budislav Vukas, *United Nations Convention on the Law of the Sea and the Polar Marine Environment*, Davor Vidas (ed.), *Protecting the Polar Marine Environment: Law and Policy for Pollution Prevention*, Cambridge University Press 2000, pp. 35–37. Note that, the question of whether or not the UNCLOS is applicable to the Arctic seas has been raised several times, due to the specific geographical, climatic, historical and political circumstances that characterize the Arctic region. It is also a fact that the UNCLOS does not indicate which sea or ocean is or isn't applicable.

[3] *See* James Graff, Fight for the top of the World, *Time Magazine*, Vol. 170, No. 13, October 2007, and Mckenzie Funk, Arctic Landgrab, *National Geographic*, Vol. 215, No. 5, May 2005.

[4] *Supra* note 1, p. 36.

[5] *United Nations Convention on the Law of the Sea*, Montego Bay, 10 December 1982, available at; http://www.un.org/depts/los/convention_agreements/texts/unclos/unclos_e.pdf (date accessed 2 February 2012).

which the Arctic may be pertinent to has been comprised in Articles 194(3) (c) and 194(5). The wordings "for preventing accidents" in Articles 194(3) (b) and 196(1)[6] highlights the notion of accidental discharge as a part of vessel-source pollution. Then again, Article 194 itself is central to any analysis of State obligation under UNCLOS to alleviate climate change as it provides the basis of certain guidance on what a State is expected to do to protect the fragile ecosystem. Moreover, Article 194(5), which deals with vulnerable seas, is of particular relevance to polar oceans.[7] Although the scope of Article 195 is obscure, it does to a certain extent broaden the concept of mitigation measures insofar as it must be designed as not to result in other environmental damage, a subject of considerable controversy in climate change.[8] Compliments may be rendered to UNCLOS for its farsightedness for providing a holistic approach to address environmental issues. These provisions although demanding further legal interpretation, substantiates the general foundation for the remainder of the provisions comprising the all-encompassing structure of prescriptive and enforcement jurisdictions.[9]

1.1.2 Flag State Jurisdiction

Flag State prescriptive jurisdiction consolidated in Article 211(2) acts as a foundation of binding obligation for all Flag States to constitute laws and regulations to adequately safeguard the marine environment by restraining vessel-source pollution. Contained in this Article with reference to those set of laws and regulations so assumed, the Flag State must ascertain that they have the parallel effect of "generally accepted international rules and standards"[10] demonstrated via "competent

[6] *Ibid. Note that,* Article 196(1) read together with Article 1(4) gives an amalgamated understanding of accidental-discharge. Article 1(4) reads thus, "…the introduction by man, directly or indirectly, of substances or energy into the marine environment, including estuaries, which results or is likely to result in such deleterious effects as harm to living resources and marine life, hazards to human health, hindrance to marine activities, including fishing and other legitimate uses of the sea, impairment of quality for use of sea and water and reduction of amenities".

[7] *Supra* note 1, p. 42, where it is stated that measures taken in accordance with Part XII "shall include those necessary protect and preserve rare or fragile ecosystems as well as the habitat of depleted, threatened or endangered species and other forms of marine life".

[8] Meinhard Doelle, *From Hot Air to Action? Climate Change, Compliance and the Future of International Environmental Law*, 2005, Thomson/Carswell, p. 23.

[9] Alan E. Boyle, "Marine Pollution under the Law of the Sea Convention", *The American Journal of International Law*, 1985, Vol. 79, p. 350.

[10] *Supra* note 3, *Note that* UNCLOS does not contain technical requirements but by means of "rules of reference" requires Flag States to give effect to existing yet generally accepted international rules and standards, i.e. GAIRAS.

international organization or general diplomatic conference".[11] The notion behind this endorsement lies in the fact that this "rule of reference" not only offers consistency, but also provides a sense of adaptability since IMO norms comprised in conventions may be sporadically revised (for updating) through an implicit acceptance amendment procedure. Article 212 (which relates to "pollution from or through the atmosphere") is another provision, although not drafted with climate change in mind, can now reasonably be clarified to apply to the prescriptive jurisdiction of the Flag State.

The inevitable obligation to ensure enforcement of international-analogous laws and regulations is integral to the prescriptive jurisdiction of a Flag State. Self-explanatory by the title, Article 217 imposes an obligation on the Flag States to ensure the implementation of national and international laws and regulations and the compliance of such norms in whichever maritime zone they might be located. Then again, Flag States are charged with conformity to generally accepted international procedure, practices and enforcement (Article 94(5)), investigating reports by States that believe that the Flag State jurisdiction and control is defective and deficient (Article 94(6)) and investigating casualties to ships flying their flags (Article 94(7)).[12] In this context, Article 222 empowers the Flag States to take necessary measures and incorporate international regulations relating to the protection of the marine environment from pollution through the atmosphere. These Flag State duties and responsibilities are tersely summarised in the Code for the Implementation of Mandatory IMO instruments (IMO Resolution A.973(24)) that gives reference to UNCLOS provisions.[13]

1.1.3 Coastal State Jurisdiction

Within internal waters Coastal State enjoys, under its sovereignty, prescriptive jurisdiction for establishing particular requirements as a condition for the entry of foreign vessels subject to due publicity to such requirements and communication

[11] *See* Report of the United Nations Secretary General, "Impact of the entry into force of the 1982 United Nations Convention on the Law of the Sea on related, existing, and proposed instruments and programmes", UN Doc. A/52/491, Section J, paras 8–9. *Note that* the main GAIRAS relating to vessel-source pollution are embodied in the global regulatory instrument adopted by IMO. Since the provision does not express the content of such laws and regulations adopted by the Flag State, this part can discretionally apply to vessels registered in their territory or flying their flags higher standard than the GAIRAS, i.e. the annexes of MARPOL 73/78 that have entered into force.

[12] John N.K. Mansell, *Flag State Responsibility: Historical Development and Contemporary Issues*, 2009, Springer, pp. 4–5.

[13] *Ibid.*, *Note that* in para 1.5, it is indicated that the central hypothesis of Flag State responsibility is that the extant regulatory regime is adequate in law, however, its enforcement does not deliver the intent of UNCLOS. A possible framework for national legislation to give effect to the relevant provisions of necessary IMO instruments is referred to in "Guidelines for Maritime Legislation", a United Nations' Publication, ST/ESCAP/1076.

of the same to competent international organization (Articles 2(1) and 211(3)).[14] An interest as between the Flag State and the Coastal State seems to be balanced by rendering Coastal State the jurisdiction to prescribe regulations concerning vessel-source pollution and by limiting the absolute freedom of navigation of Flag State(s). However, in exercising sovereignty within the territorial sea by establishing particular requirements corresponding to preservation of the environment of the Coast within that frame of jurisdiction, the Coastal State is under an obligation not to hamper innocent passage of foreign vessels (Articles 21(1) (f) and 211(4)). In this regard a significant comparison with the Flag State jurisdiction may be drawn from the fact that, the Coastal State is under no obligation to comply with GAIRAS or any minimum set of standards. On the other hand, the Coastal States' prescriptive jurisdiction is not so flexible in the innocent passage regime as regards straits where transit passage is available. Those States bordering straits may adopt laws and regulations relating to transit passage by designating sea-lanes and traffic separation schemes (Article 42(1) (a)), give effect to GAIRAS regarding discharge of oil, oily wastes and other noxious substances (Article 42(1) (b)), not apply discriminatory rules (Article 42(2)) and give due publicity to all such laws and regulations (Article 42(3)).[15] But the Convention is silent as to the consequences of lack of compliance with these standards.[16] This constrained jurisdiction accounts for, in part, both Canada's and Russian Federation's dynamic protest to the application of the straits regime in parts of their respective NWP and NSR.[17] The prescriptive jurisdiction of the Coastal State extends to EEZ where adoption of laws and regulations conforming to and giving due effect to GAIRAS is encouraged by UNCLOS (Article 211(5)).[18] Then again, a distinct reference to Arctic maybe extracted from Article 211(6) where the Coastal State is empowered to adopt additional norms in respect of EEZ if the rules and standards foreseen in Article 211(1) are insufficient and gives rise to the need to incorporate special mandatory measures

---

[14] *Supra* note 2, *Note that* these provisions can be invoked to uphold both Canada's and Russian Federation's restraining laws and regulations for control of vessel-source pollution in some areas of NWP and NSR.

[15] Veronica Frank, *The European Community and the Marine Environmental Protection in the International Law of the Sea: Implementing Global Obligations in the Regional Level*, 2007, Martinus Nijhoff Publishers, pp. 198–199, where it is highlighted that the environmental jurisdiction of Coastal States diminishes to a considerable extent in straits used for international navigation, such as the English channel, the Dover and Gibraltar straits, where foreign ships enjoy the right of innocent passage. This right, unlike the right of innocent passage in the territorial sea, can never be suspended and shall not be impeded unless there is an alternative route of similar convenience. The author goes on to say that, the regulatory powers of the Coastal States bordering the straits are limited to the prescription of navigational rules which have to conform to "applicable" international rules and need to be approved by IMO. In the strait, therefore, navigational and discharge standards contained in instruments to which Coastal States are contracting parties represent maximum standards.

[16] *Ibid.*

[17] *Supra* note 5.

[18] *Supra* note 3.

for vessel-source pollution. The reason for this reference lies in the precise wordings, "oceanographical" and "ecological condition". Nevertheless, the spontaneity of the Coastal State in consolidating such "mandatory" rules has been limited by subjecting it to consultation with any other States concerned via competent international organization. This portrays the significant effort of UNCLOS in determining a balance (between Coastal State and Flag State) so that the mandatory rules do not hamper international navigation.

The enforcement jurisdiction of Coastal States as regards territorial sea is embedded in Article 220(2) where the Coastal State may take up enforcement measures, i.e. physical inspection of the vessel, institute proceedings and detain the vessel subject to the condition that it has sufficient "clear grounds", which paves it to believe there is a certain violation of laws and regulations adopted in accordance with UNCLOS or "applicable international rules and standards".[19] This enforcement jurisdiction must, however, observe the restrictions imposed in Part II, Section 3 (innocent passage in the territorial sea) and Part XII, Section 7 (safeguards) which are intended to ensure in exercising power, States do not endanger the safety of navigation, expose the marine environment to unreasonable risk or affect the commercial interest of ships.[20] However, it is not clear what this broad provision entails in practice, more specifically with reference to the Arctic. In straits, the Coastal States may enforce appropriate measures when a vessel in transit has breached the applicable anti-pollution rules and where such breach constitutes a threat to cause "major" damage (Article 233). Most of the enforcement mechanisms available to the Coastal State in the EEZ relate to the violation of discharge or navigational standards.[21] The availability of these enforcement mechanisms, moreover, is conditional upon the gravity of discharge and little can be done before the pollution occurs.[22] Coastal States may physically inspect a foreign ship in transit pursuant to having "clear grounds for believing" that during the passage the ship has committed a breach of international anti-pollution standards resulting in "substantial discharge" causing or threatening "significant pollution" of the marine environment (Article 220(5)). However, UNCLOS does not make any distinct differences between "significant pollution" justifying an inspection and "major damage" justifying proceedings.[23] Where there has been no such infraction or discharge, the Coastal State may only request for supplementing necessary information to determine whether an infraction has taken place (Article 220(3)). Then again, Article 297(1) (b) permits Coastal States to commence procedures against Flag States, whereby in exercising the right of navigation, it has

[19] *Ibid.*

[20] *Supra* note 12.

[21] E.J. Molenaar, Book Review of "Coastal State Regulation of International Shipping" by L.S. Johnson, 22 *International Journal of Marine and Coastal Law*, 2007, pp. 183–186.

[22] *Supra* note 12.

[23] *Ibid.*

acted in contravention of UNCLOS or national and international standards espoused in conformity with this convention.

1.1.4 Lex Specialis Regime of UNCLOS

One of the deviations to Article 211(5) as regards Coastal State prescriptive jurisdiction is embodied in Article 234, where stringent legislations (stripped off the cliches of GAIRAS) are encouraged to be adopted in EEZ ice-covered waters. The provision reads thus:

> [C]oastal States have the right to adopt and enforce non-discriminatory laws and regulations [...] in ice-covered areas within the limits of the exclusive economic zone, where particularly severe climatic conditions and the presence of ice-covering such areas [...]. Such laws and regulations shall have due regard to the navigation and the protection and preservation of the marine environment based on the best available scientific evidence.[24]

Arctic marine environment from an international perspective sparks from the singular yet unparallel jurisdiction that Coastal States bordering ice-covered areas enjoy as a result of the provision negotiated as *lex specialis* at the UNCLOS. As it is considered to be a pristine eco-system predicted to be embraced by vessel-source pollution, inevitable due to the consequences of climatic changes, the Arctic Coastal States are legally ranked in pioneering development and enforcement strategies in their EEZ. The international disposition of shipping hence creates a specific nexus between IMO's global functional yet regulatory role and the special legislative and enforcement jurisdiction that Arctic Coastal States revel in the EEZ.[25]

1.1.4.1 Development of Article 234 in the Wake of Canadian National Legislation

The development of Article 234 dates back to 1970 when Canada advocated for a radical shift in the international regime of the sea to provide a license for the exercise of an extensive national legislative and enforcement jurisdiction over the global transboundary shipping activities in the Arctic via Arctic Waters Pollution

---

[24] *Supra* note 5, Article 234.

[25] Aldo Chircop, "International Arctic Shipping: Towards Strategic Scaling-Up of Marine Environmental Protection" in *Changes in the Arctic Environment and the Law of the Sea*, Myron H. Nordquist, John Norton Moore and Tomas H. Heidar (eds.), 2010, IDC Publishers, Martinus Nijhoff Publishers and VSP, p. 177, where it is indicated that it is important to acknowledge the provisions incorporated exclusively for ice-covered areas in UNCLOS and its relationship to the IMO's mandate for establishing global rules and standards for International shipping since only the IMO can espouse global rules and standards for shipping.

Prevention Act.[26,27] This was a rejoinder to the voyage of the *S.S Manhattan* tanker in the NWP and invited objections from the U.S. as regards its acquiescence with international law.[28] This contentious issue was controverted by Canada at UNCLOS where it was contrived to warrant the adoption of a provision regarding ice-covered areas pertaining to its vital interests, respectively, the affirmation of the Arctic Waters Pollution Prevention Act 1970, and the amplification of jurisdiction over its Arctic waters and NWP.[29] This was as opposed to the compromise brokered by the U.S. Delegation to UNCLOS, which layered Article 234 jurisdiction over the Coastal State's environmental jurisdiction in the EEZ.[30] The U.S. has consistently declined to accept the validity of this legislation. It seems that much of the air of conflict surrounding Arctic shipping stems from the propensity of both Canada and the U.S. to formulate issues relating to this matter in jurisdictional terms which broadly involves the demarcation of boundaries within which authority may be exercised.[31] The Canadian *lex specialis* provision had as its objective international recognition of the jurisdiction pledged by Canada under its legislation and the final and approved text of Article 234 embodied that recognition.

1.1.4.2 Pragmatic Applicability of Article 234

It is by now transparent that Article 234 basically entails that the laws and regulations adopted by the Coastal State under this provision can be more rigid than GAIRAS insofar as it lays down the Coastal State's prescriptive and enforcement

[26] *Arctic Waters Pollution Prevention Act*, R.S.C. 1985, c. A-12, adopted in 1970. Regulations were also adopted in 1978 under the Act, namely *Arctic Shipping Pollution Prevention Regulations*, C.R.C., c. 353 and *Arctic Waters Pollution Prevention Regulations*, C.R.C., c. 354. The Act is said to have extended Canadian jurisdiction to 100 nautical miles from the territorial sea baselines in waters north of 60 degrees latitude. Bill C-3 was introduced in December 2008 to extend that limit to 200 nautical miles. The Bill became *An Act to amend the Arctic Waters Pollution Prevention Act*, S.C. 2009, C. 11, proclaimed into force with effect on 1 August 2009.

[27] *Supra* note 12, p. 182.

[28] Ashley J. Roach and Robert W. Smith, *United States Responses to Excessive Maritime Claims,* Martinus Nijhoff Publishers (The Hague/Boston/London), 1992, Second edition, pp. 339–353, *See inter alios* D.M. McRae and D.J. Goundrey, "Environmental Jurisdiction in arctic waters: the extent of article 234", *University of British Columbia Law Review,* Vol. 16:2, 1982, and R. Hubert, "Article 234 and Marine Pollution Jurisdiction in the Arctic" in Oude Elferink A.G. and Rothwell, D.R. (eds.), *The Law of the Sea and Polar Maritime Delimitation and Jurisdiction,* 2001, Martinus Nijhoff Publishers, pp. 249–267.

[29] D.M. McRae, "The Negotiation of Article 234" in *Politics of the Northwest Passage*, Franklyn Griffiths (eds.), 1987, McGill-Queen's University Press, pp. 98–114.

[30] *Supra* note 12.

[31] Oran R. Young, "Arctic Shipping: An American Perspective" in *Politics of the Northwest Passage*, Franklyn Griffiths (eds.), 1987, McGill-Queen's University Press, p. 124.

jurisdiction.[32] Although the *prima facie* virtue of this provision is an approximative ascription of jurisdictional power, by dint of certain wordings[33] the influence of this provision has been restricted in the event of implementation. It also implies that the principles of reason and logic must follow as not to interfere with international navigation and to have due regard to protect and preserve the marine environment based on what is termed as "best available scientific evidence". Hence, Article 234 does not provide Coastal States of the Arctic region with a *carte blanche* regulatory dominion.[34]

In reality, Article 234 has been termed as "probably the most ambiguous, if not controversial clause, in the entire treaty".[35] Interpretation and lucid applicability of Article 234 via *travaux préparatoires* is cumbersome.[36] The provision comprises that the regulatory authority must be exercised within the limits of EEZ. In some instances this includes the territorial sea and in others it does not, thus limiting its application to the EEZ.[37] Then again, among other issues that have been highlighted by the international community are unilateral stricter CDEM standards, transit fees, and obligatory ice-breaker escort and other measures that may hinder smooth navigation. In addition, the provision suggests that the regulation has to be for environmental protection purpose. However, if the regulation is exclusively for safety objectives, there is a chance that Article 234 may not cover that situation and ultimately it may be difficult to distinguish between the two purposes of regulation.[38] In either case, Article 234 must be based on the best approved scientific substantiation. Moreover, with the impact of climate change the presence of ice in such areas "for most of the year" is no longer a reality. A question is left on the face of this provision as to whether the dramatic deterioration of sea-ice shall prove the inapplicability of Article 234 in the coming decades. It seems that the requirement for Arctic Coastal States to approve laws and regulations through IMO is absent. It is believed that the policy-makers' efforts served the best interest for those Arctic Coastal States by allowing them to adopt higher national standards

---

[32] Alan Khee-Jin Tan, *Vessel-Source Marine Pollution: The Law and Politics of International Regulation,* Cambridge University Press, 2006, p. 234.

[33] *Supra* note 3, Article 234, *Note* the wordings "non-discriminatory", "shall have due regard" and "based on the best available".

[34] Aldo Chircop, "Challenges for the Regulation of International Shipping through the Arctic" in *Impacts of Climate Change on the Maritime Industry*, The proceedings of the Conference on Impacts of Climate Change on the Maritime Industry, 2–4 June 2008 (Sweden), Neil Bellefontaine and Olof Linden (eds.), 2009, World Maritime University Publications, p. 212.

[35] Cynthia Lamson, "Arctic Shipping, Marine Safety and Environmental Protection", *Marine Policy*, Vol. 11, 1987, p. 3, and Øystein Jensen, *The IMO Guidelines for Ships Operating in Arctic Ice-Covered Waters: From Voluntary to Mandatory Tool for Navigation Safety and Environmental Protection?,* The Fritjof Nansen Institute, FNI-rapport 2/2007, p. 7.

[36] Myron H. Nordquist, *United Nations Convention on the Law of the Sea, 1982: A Commentary,* University of Virginia, Center for Oceans Law and Policy, Vol. 3, pp. 392–398.

[37] *Supra* note 12.

[38] *Ibid.*

as opposed to negotiated standards.[39] What started out as a negotiation provision now forms a part of the larger package deal of UNCLOS and would not be an aberration if non-Arctic States expressed an interest in this provision. Till date, Canada and the Russian Federation have implemented this provision. The relationship between Canadian regulation with a view to protecting the Arctic marine environment and Canadian undertakings assumed under MARPOL 73/78 is articulated in a reservation to the latter treaty, using the authority of Article 234.[40] The Russian claim to jurisdiction over the NSR is based on Article 234 and under the Russian Regulations, all vessels intending to enter the NSR should give advance notifications to Russian authorities and submit therewith an application for guiding (which implies paying a fee for using the Route).[41] Diplomatic endeavours between involved parties (especially between Canada, the U.S. and the Russian Federation) have had a certain degree of success and climate change acting as a catalyst, time will only reflect the reaction of future navigators (of the Arctic frontier) and concerned State parties.

1.2 Analysis of IMO Instruments Pertaining to Arctic

It is estimated that most of the international instruments which correspond to the protection of the marine environment are applicable in Arctic waters.[42] While UNCLOS provides the general jurisdictional provisions for ice-covered areas, IMO has a bearing role in the operationalisation of these provisions.[43] The varying requirements among states in terms of CDEM, emission standards and ship navigation have been sought to being harmonised by IMO in view of the commercial, shipping and environmental aspects inherent in international shipping. Pollution

---

[39] *Ibid.*

[40] *Ibid.*, where it has been highlighted that "Consequently, Canada considers that its accession to the Protocol of 1978, as amended, relating to the International Convention for the Prevention of Pollution from Ships, 1973 (MARPOL 73/78) is without prejudice to such Canadian laws and regulations as are now or may in the future be established in respect of Arctic waters within or adjacent to Canada (IMO, 2005)".

[41] Katarzyna Zysk, *Russia's Arctic Strategy: Ambitions and Constraints*, Joint Force Quarterly, Issue 57, 2nd Quarter, 2010, available at; http://www.ndu.edu/press/lib/images/jfq-57/zysk.pdf (date accessed 20 February 2012), where the contemporary practice has been highlighted by stating that, the question of the legal status of the NSR complicates the fact that it is not a single shipping channel, but a series of different shipping lanes stretching between 2,200 and 2,900 nautical miles, depending on ice conditions. According to Russian experts, "the integral nature of the NSR as a transport route is not affected by the fact that individual portions of it, at one time or another, may pass outside boundaries of internal waters, territorial waters and EEZ, i.e. it may pass into the high seas." The NSR may thus include sea lanes running beyond Russia's EEZ as long as part of the voyage includes waters under undisputed Russian jurisdiction.

[42] Donald R. Rothwell, *The Polar Regions and the Development of International law*, 1996, Cambridge University Press, p. 213.

[43] *See Supra* note 6 for further analysis of IMO's role.

from ships, when it is not accidental, is operational in nature and emanates from a routine of the manner in which the ship operates.[44] Under the regulatory regime of IMO, MARPOL 73/78 which is the most relevant IMO instrument deals with regulating the discharge and emission standards. Moreover, for ice-covered areas it is significant to acknowledge the Guidelines for Ships Operating in Arctic Ice-Covered Areas (here and after, referred to as Polar Shipping Guidelines).[45]

1.2.1 Regulations Governing Operational Discharge Under MARPOL 73/78

MARPOL 73/78 is not limited only to the regulation of oil pollution from ships but also comprises various types of ship-source pollution insofar as it supplements evidence of internationally agreed standards of environmentally sound management for the transport of hazardous wastes and chemicals.[46] Given its scope and applicability to all vessels flying their flags or under the authority of State party, it contains in its Annexes restrictions as regards voluntary discharge from those vessels.[47] These rules and standards for the Arctic have been termed as "Level 1" standard of protection which is the generally accepted minimum under MARPOL 73/78. Annexes I, II and V pave the way for the possibility to establish areas designated as "Special Areas" and "Sox emission control areas" where the specific sensitivity justifies the application of more stringent discharge and emission standards. The number of ratifications[48] of MARPOL 73/78 Annexes by Arctic countries refers to the efforts and assiduity to harmonize discharge and emission standards. However, the discharge of oil from machinery spaces of all ships, which is regulated by

[44] Proshanto K. Mukherjee, "The Penal Law of Ship Source Marine Pollution" in *Law of the Sea, Environmental Law and Settlement of Disputes: Liber Amicorum Judge Thomas Mensah*, 2007, Martinus Nijhoff Publishers, pp. 480–485, *Note* that marine pollution in this context has been immaculately divided into "voluntary" and "accidental" and "voluntary" has been further sub-divided into "deliberate (dumping)" and "operational (discharge)".

[45] *Guidelines for Ships Operating in Arctic Ice-Covered Areas*, Adopted by IMO MSC/Circ. 1056, MEPC/Circ. 399, of 23 December 2002, available at http://www.imo.org/blast/blastDataHelper.asp?data_id=29985&filename=A1024(26).pdf (date accessed 22 February 2012).

[46] Patricia Birnie and Alan Boyle, *International Law and the Environment*, 2002, Oxford University Press, p. 363, *Note* that Article 2 of MARPOL 73/78 defines "discharge" which covers any means of release such as disposal, spilling, leaking, pumping, emitting and emptying, but does not include dumping in the meaning of the London Convention.

[47] Whereby, Annex I deals with prevention of pollution by oil, Annex II on control of pollution by noxious liquid substances in bulk, Annex IV on prevention of pollution by sewage from ships, Annex V on prevention of pollution by garbage from ships and Annex VI on the prevention of air pollution from ships.

[48] All the Arctic countries are signatories to MARPOL 73/78.

regulation 15 (Annex I), restricts the discharge of any amount in the Antarctic from ships less than 400 Gross Tonnage unless it complies with Regulation 15-C, but does not provide special attention in that regard to the Arctic Ocean. Similarly, Annex II (Discharge of Noxious Liquid Substance) and Annex V (Disposal of Garbage) does not consider Arctic to be a part of "Special Areas" as opposed to the Antarctic where operational discharges are unauthorized. Whereas Canada and the Russian Federation have adopted stringent regulation under the terms of Article 234 of the UNCLOS in the Arctic north of 60 degrees North latitude, Canada has deliberately precluded MARPOL 73/78 for those areas.[49] In addition to the provisions and standards of MARPOL 73/78 applicable in ice-covered areas, those States demand for the compliance of stricter standards. Unlike the Antarctic, there is a vacuum of designated areas, i.e. "Sox emission control area" for the purpose of special protection under MARPOL 73/78. These nominated areas would, to a greater extent, strengthen the protection of the marine environment in Arctic ice-covered areas specifically in those lying in areas beyond national jurisdiction.

1.2.2 Investigating the IMO Polar Shipping Guidelines

The significant initiative for regulating ship construction, equipping and operations in polar waters took place under the auspices of IMO instigated by the disaster of *Exxon Valdez*, off the coast of Alaska in 1990.[50] IMO noted that the existing Arctic national legislation as regards technical requirements for navigation within the EEZ of Canada, Norway, Russia and the U.S. differed extensively, making it impossible for the Flag State to comply with every law in the course of one and the same voyage.[51] Hence, the Polar Shipping Guidelines evolve from an integrated approach of IMO which originally revolved around promoting standards for safety of navigation. Although the history of the Polar Shipping Guidelines date back to 1991, when Germany proposed the inclusion of a provision which entailed that ships intended for service in the polar regions should have suitable ice strengthening for polar conditions, it was in fact Canada on behalf of Outside Working Group that submitted the international Code of Safety for Ships in Polar Waters in 1998. This Code after a certain extent of revision and exclusion of Antarctica developed as the Polar Code by the Marine Environment Protection Committee at its 76th session (December 2002).[52]

---

49 *Bill C-3: An Act to Amend the Arctic Waters Pollution Prevention Act*, Legislative Summary, LS-617E, Penny Becklumb Industry, Infrastructure and Resources Division, 19 December 2008, Accessed at http://www.parl.gc.ca/Content/LOP/LegislativeSummaries/40/1/401c3-e.pdf (date accessed 22 February 2012).

50 *Supra* note 32, p. 8.

51 *Ibid.*

52 *Ibid.*

It is already acknowledged that there is a dearth of waste reception facility in the Arctic and the polar environment coupled with the special navigation criteria imposes additional demands on the shipping industry than other general standards prescribed by IMO. The Guidelines relate to ships operating in Arctic ice-covered waters as defined in para G-3.2, and while engaged in International voyages.[53] The environmental protection aspect is embedded in Part-D (Chapter 16) of the Polar Shipping Guidelines.[54] This part is divided into three sections where Section 16.1 deals with the general provisions, Section 16.2 comprises the prescriptive part relevant to equipment and materials and Section 16.3 directly deals with the protection of the environment which restrains itself to remit to national and international rules regulating discharges and emissions from ships but affixes no specific standards which take in consideration the particular effects of vessel-source intentional pollution in Arctic ice-covered areas.[55] Acknowledging the fact that the Arctic lacks "Special areas" or "Sox emission control areas" under MARPOL 73/78, the Polar Shipping Guidelines with this simple remission as incorporated in Chapter 16 adds no particular reinforcement to the protection of the Arctic marine environment and this vacuum till date is consistent. Then again, the Guidelines' non-legally binding nature and limited regulation on environmental protection set no obligation on the States in respect to operational vessel-source pollution and special particularities in ice-covered areas.

1.3 Deliberate Dumping in Arctic and International Law

The LOSC is appropriate in a general, rather than a precise sense to address ocean dumping issues in the Arctic and sets forth a global framework for ocean dumping.[56] The provisions of UNCLOS which relate to pollution from ocean dumping are Articles 1(1) (5), 210 and 216 where Article 210 presumably refers to LC72[57] and its Annexes indicating the fact that national regulations shall be no less operative than the rules and standards set globally. LC72 is a pertinent and

[53] *Ibid.*

[54] *Ibid.*

[55] *Ibid.*

[56] *PAME: Working Group on the Protection of the Arctic Marine Environment*, Report to the Third Ministerial Conference on the Protection of the Arctic Environment, 20–21 March 1996 Inuvik, Canada, Ministry of Environment, p. 95, *Note* that the Convention provides minimum standards for contracting parties in that domestic laws, regulations and measures shall be no less effective in preventing, reducing and controlling pollution from ocean dumping than the global rules and standards. *See* also Olav Schram Stokke, *Radioactive waste in the Barents and Kara Seas: Russian implementation of the global dumping regime* in Protecting the Polar Marine Environment, Davor Vidas (ed.), 2000, Cambridge University Press, p. 203.

[57] Adopted on 29 December 1972 in London, Mexico City, Moscow and Washington, D.C., and entered into force on 30 August 1975.

comprehensive instrument which consists of three Annexes[58] to address ocean dumping in the Arctic which relates to ocean dumping and incineration. The five Arctic countries are parties to the LC72 and have implemented it domestically, whereby the Russian Federation and the U.S. are not parties to the protocol of 1996.[59] Canada fulfils its international obligations, in part, through Part 7, Division 3 (Disposal at Sea) of the Canadian Environmental Protection Act, 1999. The U.S. has adopted Ocean Dumping Act, codified as Titles I and II of the Marine Protection, Research and Sanctuaries Act of 1972, 33 U.S.C. (paras 1401 et seq.). Then again, Sweden's Law 1971:1154 on Prohibition of Dumping on Wastes at Sea (pp. 1–3) is a conforming national law in this regard. LC72 is applicable to all marine waters outside internal waters and sets a minimum standard for all States on the basis of categories of pollutants and a system of permits for those substances permissible for dumping.[60] Although LC72 has displayed competency to accommodate new expansion through these flexible Annex structures, it lacks a formal non-compliance procedure and no action has been taken to improve liability procedures. Non-acceptance of LC72 ban on low level radioactive waste dumping by the Russian Federation portrays a substantial gap in conformability relevant to protecting the Arctic environment although the Russian Federation's Federal State Programme has implemented national regulation.[61] On 7 November 1996, a Protocol to the Convention was adopted which was intended to gradually replace LC72.[62] The Protocol explicitly highlights the precautionary principle, requiring that the appropriate preventative measures ought to be taken when there are reasons to believe that wastes or other matter introduced into the marine environment are likely to cause harm, even when there is no conclusive evidence to verify a causal relation between inputs and their effects and goes on to say that "the polluter should, in principle, bear the cost of pollution". The

[58] Dumping of matter listed in Annex I is prohibited; dumping of matter listed in Annex II is allowable only by special permit; dumping of matter listed in Annex III is allowable only by general permit.

[59] Official webpage of IMO, available at; http://www.imo.org/About/Conventions/StatusOfConventions/Pages/Default.aspx (date accessed 29 February 2012).

[60] *Ibid.*

[61] Official Webpage of IMO, available at; http://www.imo.org/About/Conventions/StatusOfConventions/Pages/Default.aspx (date accessed 23 February 2012), *Note* that the information gathered from the webpage was as regards the status on LC72.

[62] Official Webpage of IMO, available at; http://www.imo.org/OurWork/Environment/SpecialProgrammesAndInitiatives/Pages/London-Convention-and-Protocol.aspx (date accessed 23 February 2012), *Note* that the information fathered from the Webpage was as regards the summary of LC72.

Protocol was amended in 2006[63] to accommodate a basis in International environmental law for the regulation of carbon capture and storage under the seabed, for purposes of tackling climate change and ocean acidification. This amendment is a big step for LC72 to mitigate complex yet technical environmental issues in the Arctic region.

1.4 International Legal Regime of Accidental Pollution in the Arctic

Oil-spill, more specifically, accidental oil-spill in ice-infested waters could generate severe marine pollution consequences and a great risk of detriment to the environment for prolonged periods of time. Instances may be drawn from the double-purpose passenger and supply ship *Bahia Paraiso*, which grounded in January 1989 in the Antarctic, exemplifies the dangerous effects of pollution as a result of increased shipping traffic in ice-areas. On 24 March 1989, the tanker *Exxon Valdez* spilled 11 million gallons of crude oil into the Pacific Gulf of Alaska and over 1,200 miles of coastline of the Alaska Peninsula were contaminated with oil, which caused massive damage to the natural marine environment in addition to other damages.[64] Under OPRC[65] and OPRC-HNS Protocol,[66] governments of Coastal States have been prescribed to establish measures for dealing with accidental pollution, either nationally or via bilateral cooperation.[67] Parties to the Convention are also required to provide assistance to others in the event of a pollution emergency and provision is made for the reimbursement of any assistance provided. It is significant to note that this Convention provides for the IMO to play an important coordinating role.[68] With reference to the Arctic, the Costal States are dependent on MOU and regional or bilateral arrangements are in place that provide

63 The amendments entered into force in February 2007.

64 James D. Ford, Barry Smit, Johanna Wandel, "Vulnerability to Climate Change in the Arctic: A case study from Arctic Bay, Canada" in *Global Environmental Change*, 2006, Vol. 6, Issue: 2, Elsevier publisher, pp. 145–160.

65 (IMO) Adoption: November 1990; Entry into force: 13 May 1995.

66 (IMO) Adopted in 2000.

67 Official Webpage of IMO, available at; http://www.imo.org/About/Conventions/ListOfConventions/Pages/Protocol-on-Preparedness,-Response-and-Co-operation-to-pollution-Incidents-by-Hazardous-and-Noxious-Substances-(OPRC-HNS-Pr.aspx (date accessed 23 February 2012).

68 *Ibid.*

a framework for cooperation among Arctic States under OPRC.[69] Several Arctic states have joint contingency planning arrangements and they include, among others, the Canada/United States Joint Marine Pollution Contingency Plan for the Beaufort Sea area, the Russia/USA Joint Marine Pollution Contingency Plan, the joint Russian/Norwegian Plan for the Combating of Oil Pollution in the Barents Sea and the Canada/Denmark Agreement for Marine Environmental Cooperation, which includes annexes for responding to shipping and offshore hydrocarbon spills.[70] The other international regime which deals with accidental pollution is Annex I and Annex II of MARPOL 73/78. Annex I covers prevention of pollution by oil from accidental discharges and the 1992 amendments to Annex I made it mandatory for new oil tankers to have double hulls, which was subsequently revised in 2001 and 2003.[71] Then again, Regulation 13F relating to "Prevention of oil pollution in the event of collision or stranding" does not provide specific

---

[69] "Governance of Arctic Shipping" in *Arctic Marine Shipping Assessment*, 29 April 2009, available at; http://www.arctic.gov/publications/AMSA/governance.pdf (date accessed 23 February 2012). Note that OPRC is a framework for international cooperation in combating incidents or threats of marine oil pollution, to which all five Arctic States including the other three States that are located in the Arctic circle, i.e. eight Arctic States are parties. Article 10 of the OPRC promotes the development of bilateral and multilateral agreements for oil pollution preparedness and response, such as on a regional basis, *See* also, *See Infra* note 33 in Chap. 2, where it is elaborated that, the OPRC convention provides yet another foundation for Polar States to develop comprehensive strategies to respond to maritime incidents, which have an environmental repercussion. By placing minimum obligation on the Flag States to ensure that their vessels have an oil pollution emergency response plan. Polar States can be reassured that the vessels from OPRC convention parties, which operate in their waters, have met this standard. The requirements for cooperation among States to deal with such incidents are also important, given the difficulties in responding to emergencies in polar waters. Nevertheless, the significant issue is whether States should meet higher standards than the OPRC when venturing in the Arctic waters. The Canadian AWPPA is a notable example in this regard.

[70] *Ibid.*

[71] Official Webpage of IMO, available at; http://www.imo.org/about/conventions/listofconventions/pages/international-convention-for-the-prevention-of-pollution-from-ships-(marpol).aspx (date accessed 23 February 2012), parties to MARPOL 73/78 may enforce the convention in three ways; through ship inspections to ensure vessels meet minimum technical standards, by monitoring ship compliance with discharge standards, and by punishing ships which violate the standards, *See* also Andrew Griffin, "MARPOL 73/78 and Vessel Pollution: A Glass Half Full or Half Empty?", *Indiana Journal of Global Legal Studies*, 4 January 1994, Vol. 1, Issue 2, Article 10, pp. 489–513, *Note* that, the revised MARPOL 73/78, Annex I *Regulations for the prevention of pollution by oil* was adopted in October 2004 and enters into force on 1 January 2007. It comprises the various amendments adopted since MARPOL 73/78 entered into force in 1983, including the amended regulation 13G (regulation 20 in the revised Annex) and regulation 13H (regulation 21 in the revised Annex) on the phasing-in of double hull requirements for oil tankers. It also separates, in different chapters, the construction and equipment provisions from the operational requirements and makes clear the distinctions between the requirements for new ships and those for existing ships. Unique and updated as it may seem, the revision provides a more user friendly and compatible, simplified and transparent Annex I, to be regulated in regions which are sensitive and pristine in nature. This is, of course, based on a general context, rather than being tailor-made for a specific region including the Arctic.

reference to the Arctic which requires more stringent policies because of the vulnerable characteristics it possesses. Similar comments have been aimed against Annex II relating to control of pollution by noxious liquid, which has completely ignored the Arctic regime.

Although trans-Arctic shipping beyond the EEZ and across the high seas of the Arctic raises *governance issues*,[72] the IMO has enacted an Intervention Convention which deals with instant measures on the high seas as may be necessary to prevent, mitigate or eliminate danger to its coastline or related interests from pollution by oil or the threat thereof, following a maritime casualty.[73] Although the IHSCO has no implicit provision on the Arctic, the parties to the Convention have an option to implement intervention policies under respective NCPs. Intervention in this context is directly connected to the preparedness and response and is the defence mechanism to respond rapidly to accidental pollution. OSI is the new buzzword when it comes to dealing with accidental pollution in the high seas where Coastal States may take measures rendered necessary by the urgency of the situation, without prior notification or consultation or without continuing consultations already begun.[74] Although this is a simple exception to the provisions of "consultation", it is still important to consult with neighbouring states prior to initiation of intervention measures. "Places of Refuge", in connection to the intervention policy, is a legal theory which comprises a part of the intervention procedure where the polluting ship is towed to a nearby place to deter oil pollution in pristine waters or waters designated as SSA.[75] But problems arise when national authorities are given unlimited

[72] *N.B.* because a Coastal State's authority to regulate foreign shipping does not extend to the high seas, transiting ships would only be subject to global shipping safety, environmental and security rules and standards adopted through the IMO and as may be applied by the Flag States.

[73] International Convention Relating to Intervention on the High Seas in Cases of Oil Pollution Casualties (with annex official Russian and Spanish translations and Final Act of the International Legal Conference on marine pollution damage, 1969). Concluded at Brussels on 29 November 1969, available at; https://treaties.un.org/doc/Publication/UNTS/Volume%20970/volume-970-I-14049-English.pdf (date accessed 5 August 2014), *N.B.* that the 1969 Intervention Convention applied to casualties involving pollution by oil. In view of the increasing quantity of other substances, mainly chemical, carried by ships, some of which would, if released, cause serious hazard to the marine environment, the 1969 Brussels Conference recognized the need to extend the Convention to cover substances other than oil. The 1973 London Conference on Marine Pollution therefore adopted the Protocol relating to Intervention on the High Seas in Cases of Marine Pollution by Substances other than Oil. This extended the regime of the 1969 Intervention Convention to substances which are either listed in the Annex to the Protocol or which have characteristics substantially similar to those substances. The 1973 Protocol entered into force in 1983 and was amended in 1996 and 2002 to update the list of substances attached to it.

[74] *Ibid.*, where Article III(b) states that "… [t]he coastal State shall notify without delay the proposed measures to any persons physical or corporate known to the coastal State, or made known to it during the consultations, to have interests which can reasonably be expected to be affected by those measures. The coastal State shall take into account any views they may submit".

[75] Morrison, Anthony P. (2012) Places of Refuge for Ships in Distress: Problems and Methods of Resolution. Legal Aspects of Sustainable Development. Martinus Nijhoff. doi: 10.1163/9789004218888.

power to override international law to intervene in oil pollution incidents and select a place of refuge which might have an effect on the adjacent waters of other Coastal States. This might be done under the sphere of NCPs and respective provisions of the IHSCO where Coastal States may take action without consulting other States.[76] NCPs relating to OSI developed under the auspices of IOs may, in light of the Arctic geographical situation, invite additional conflict based on individual definition of

---

[76] Official Website of Transport Canada, available at; http://www.tc.gc.ca/eng/marinesafety/tp-menu-515-3050.htm (date accessed 5 August 2014), *N.B.* Many states, including Canada, give their ministers, harbour authorities or delegated persons the power to permit the entry, or conversely, to take unilateral action to remove or destroy a vessel where there is a risk to the safety of a port, or the coastal environment. This in other words translates into the "power of intervention" to respond to an oil-threat at the beginning of a casualty. This power underlies the importance of "places of refuge" as an international issue requiring action for the protection of commercial, social and environmental interests. The Places of Refuge Contingency Plan is already in place and applies to all situations where a ship is in need of assistance and requests a place of refuge within Canadian waters. This includes Canada's internal waters, territorial sea and the Exclusive Economic Zone. Section 3 of the Places of Refuge Contingency Plan stipulates that the most suitable place of refuge can only be determined after the details of the specific incident are known and thoroughly considered. To pre-designate places of refuge may be of limited value, as the limitations, operational considerations, hazards and associated risks will vary greatly with each incident. Experience in Canada has shown that because no two incidents, and the circumstance surrounding the incident, are very similar, the value of pre-planning lies primarily in ensuring information will be readily available (i.e. nautical charts and publications, port information, environmental and sensitivity data), along with the relevant specialists. On the other hand, the approaches adopted by the Norwegian Coastal Directorate is to identify and list "places of refuge" and places of grounding in the NCAs Emergency Response Plan. These "places of refuge" and grounding are utilized in cases where there is a danger of severe pollution as a result of accidents at sea. In other words, they form an integral part of oil spill intervention for Norway. The procedure developed by the Directorate assumes that allowing the leaking oil cargo to contaminate the sea along Norway's coastline could lead to the pollution of vast coastal areas and hence towing the damaged ship to a place of refuge or grounding would be the preferred option, as the spill could then be better controlled, contained and cleaned. The procedure assumes that the ship in need of assistance passing along Norway's coast can be towed ashore to a grounding site to prevent it from sinking. As each operation is unique, the Emergency Response Plan requires evaluation on a case-by-case basis of using a place of refuge or grounding from the list provided. This list contains 69 designated places of refuge and 62 places of grounding along the Norwegian coast. These designated sites help in the process of accurate, efficient and timely decision-making during an emergency involving a damaged vessel. The two lists are constantly reviewed and updated as new information is obtained on the sea, environment and dynamics of ship traffic. It is by now understood that the customary right of access to a place of refuge for vessels in distress has become a sensitive yet complex issue from increasingly conflicting values. The need to stop a leak or contain the damage done by oil spill might be the first step in intervention, however, determining and access to a "place of refuge" might be seen as a major part of the intervention so that every possibility of a "near spill" occurring again from the same leak might be taken care of in the so-called place of refuge. Till date, there has been numerous interventions that have taken place in various incidents, but they differ from each other in terms of ship-damage and authority-action. For some, the immediate response was prolonged because of the given situation and for some it was not possible because of late-response. Either ways, states continue to address intervention in their respective national legislation to deal with circumstances as an accountable state under international law.

"marine protection" and already existing national legislation for respective parts of the Arctic marine environment. The problem for oil spill intervention in the Arctic can be termed as complex because first and foremost, there still remains the question of geographical delimitation issues.[77] The concept of "places of refuge" might act as a catalyst to this *status quo* complexity where Arctic Coastal States might be in a situation to select a place bordering another Arctic Coastal State as a part of OSI. This might result in a conflict since the IHSCO has given unlimited authority to States involved in immediate intervention that seeks to control pollution to the Arctic environment via rapid response. If for any reason other Arctic Coastal States refuse to provide a place of refuge to the acting authority—then it might hamper the intervention in context. Moreover, the exceptions to the provisions to act without "consultation" might not be acceptable to Canada if for any reason the U.S. decides to act unilaterally without considering the impact that the wrecked vessel from the maritime casualty have on the Arctic waters of Canada. This as per observation, is a major drawback in the subject matter of OSI and needs to be clarified by IMO.

[77] *N.B.* Disputes in the field of the law of the sea cannot be ruled out, however, for example over the delimitation of the continental shelf. Several unresolved issues can be mentioned in this context: (1) The United States and Canada are involved in a dispute over the Northwest Passage and a part of the Beaufort Sea, which is estimated to hold vast oil deposits. The United States considers the Northwest Passage as an international strait, whereas Canada considers the route its internal waters. (2) Denmark and Canada on the one hand, and Russia on the other hand, disagree on jurisdiction over the Lomonosov Ridge in the Arctic Ocean. (3) Most nations reject Norway's claim of a 200 nautical miles zone around Svalbard on the basis of conditional sovereignty over the island and have refused to recognise their "fisheries protection zone" around it. (4) A dispute is ongoing between Canada and Denmark over Hans Island, which is located in the strait that separates Ellesmere Island from northern Greenland and connects Baffin Bay with the Lincoln Sea.

Chapter 2
Pertinent National Legislation of Arctic States

2.1 Canada and the Arctic

Arctic marine transportation is a central element of Canada's INS which supports the key priorities in economic, social development, environmental protection and sovereignty.[1] The Canadian government has made various commitments to defend its vision for the North and to enhance Canada's security and enforcement capabilities in the Arctic. Canada's OA declares a territorial sea of 12 nm from the territorial sea baselines, a contiguous zone adjacent to the territorial sea out to 24 nm from the territorial sea baselines, an EEZ adjacent to the territorial sea and extending out to 200 nm from the territorial sea baselines, and a continental shelf of at least 200 nm or further in case of an extended continental margin.[2] Specifically recognizing the navigational rights and freedoms of foreign vessels for each zone, the Oceans Act does comprise some provisions aimed at protecting those rights and freedoms in terms of Canada's enforcement jurisdiction.[3] The underlying principle which governs the OA is the concept of "integrated management", which is a concept linked to the aspect of good ocean governance.[4] In 2002, under the guidance

---

[1] *Infra* note 27 in Chap. 8.

[2] Oceans Act (Consolidation), S.C. 1996, c. 31, available at; http://laws-lois.justice.gc.ca/PDF/O-2.4.pdf (date accessed 5 August 2014).

[3] *Ibid.*

[4] *Ibid.*, where Section 30 of Part II states that "[t]he Minister, in collaboration with other ministers, boards and agencies of the Government of Canada, with provincial and territorial governments and with affected aboriginal organizations, coastal communities and other persons and bodies, …; shall lead and facilitate the development and implementation of plans for the integrated management of all activities or measures in or affecting estuaries, coastal waters and marine waters that form part of Canada or in which Canada has sovereign rights under international law. "Moreover, in order to implement the IMP, the Minister"… (a) shall develop and implement policies and programs with respect to matters assigned by law to the Minister; (b) shall

T. Johansson and P. Donner, *The Shipping Industry, Ocean Governance and Environmental Law in the Paradigm Shift*, SpringerBriefs in Law,
DOI 10.1007/978-3-319-12541-1_2

of the OA, the federal government developed the Oceans Strategy, which outlines a strategic path for realizing international commitments and domestic mandates for marine conservation through an ecosystem-based management approach.

2.1.1 Investigating Torrey Canyon Antecedents

The *Torrey Canyon* incident had its worst impact on the miles of beautiful, rural coast of Southwestern England.[5] It was opined that neither existing international

Footnote 4 (continued)
coordinate with other ministers, boards and agencies of the Government of Canada the implementation of policies and programs of the Government with respect to all activities or measures in or affecting coastal waters and marine waters; (c) may, on his or her own or jointly with another person or body or with another minister, board or agency of the Government of Canada, and taking into consideration the views of other ministers, boards and agencies of the Government of Canada, provincial and territorial governments and affected aboriginal organizations, coastal communities and other persons and bodies, including those bodies established under land claims agreements, (i) establish advisory or management bodies and appoint or designate, as appropriate, members of those bodies, and (ii) recognize established advisory or management bodies; and (d) may, in consultation with other ministers, boards and agencies of the Government of Canada, with provincial and territorial governments and with affected aboriginal organizations, coastal communities and other persons and bodies, including those bodies established under land claims agreements, establish marine environmental quality guidelines, objectives and criteria respecting estuaries, coastal waters and marine waters." In this regard it is also important to note the work of Transport Canada. Several federal departments and agencies share responsibility for managing and protecting the remote Arctic coastline. Transport Canada, for instance, administers the Marine Transportation Security Regulations, while the Canadian Coast Guard is the lead federal agency when a marine pollution incident occurs north of 60 degrees north latitude (hereafter "north of 60"). National Defence is responsible for Canada's National search and rescue (SAR) program, while the Coast Guard is responsible for its marine component.

[5] Albert E. Utton, *Protective Measures and the "Torrey Canyon"*, 9 B.C.L. Rev. 613 (1968), available at; http://lawdigitalcommons.bc.edu/bclr/vol9/iss3/4, (date accessed 27 February 2012, the reaction of it has been further demonstrated in p. 617 where it has been highlighted that in view of this threat of future incidents such as the *Torrey Canyon* disaster and in response to the statement in the British White Paper on the *Torrey Canyon* incident that "[t]he law relating to international shipping is … in a number of ways quite out of date," "it is appropriate to explore what measures a coastal state may take in order to protect itself from the threat of oil pollution emanating from damaged vessels off its coasts. In examining the question of *what* measures may be taken, it is necessary to answer the included question of *when—prior* or subsequent to collision—such measures may be taken and *where* in the adjacent high seas they may be taken, *See* also The Times (London), March 29, 1967, at 1, col. 1, where it is highlighted that The *Torrey Canyon* went aground on the Seven Stones Reef, outside British territorial waters, 15 miles west of the Cornish peninsula, and about 10 miles from the British Isles of Scilly and thus, measuring from the Isles of Scilly, the wreck occurred well within the 12-mile maximum for contiguous zones set by the 1958 Geneva Convention on the Territorial Sea and the Contiguous Zone. Within this contiguous zone, and outside the British territorial sea," the British took direct action. In addition to the bombing, measures used in the *Torrey Canyon* case included spraying the oil slick with emulsifying agents or coagulating agents, burning the escaping oil by use of combustible agents, and using urethane booms.

conventions nor national legislation dealing with oil pollution specifically addressed issues relating to what protective measures a Coastal State may adopt to address incidents such as the *Torrey Canyon*. In examining the history, it may be clear that the 1954 Convention for the Prevention of Pollution of the Sea by Oil, as an instance, established zones extending 50 miles from the coast in which ships were prohibited from discharging oil into the sea, and prescribed penalties to be imposed by signatory states for such discharge.[6] Similarly, national legislation of that era were aimed at punishing the routine discharge of oil, for example, the U.S. Oil Pollution Act of 1924[7] which prohibited "the discharge of oil by any method…into or upon the navigable waters of the United States…". Then again, there existed British legislation, i.e. Oil in Navigable Waters Acts of 1955[8] and 1963[9] which made the discharge of oil punishable, but specifically exempted were those instances where the escape of oil was caused by damage to the vessel. This exception was identical to and suited the *Torrey Canyon* situation. The general consensus was that there was the need for a new legal regime, while acknowledging the need for State intervention on the high seas in cases of grave emergency, clearly restricted that right to protect other legitimate interests.[10] Thus, was introduced by IMO the first International Convention related to liability and compensation, i.e. the CLC[11] and the FUND[12] Convention. There was, therefore, no clear policy during the 1950s and early 1960s on the status of the Arctic waters and, Canada's response to the convention promulgated in 1969 could not be determined as positive.

2.1.2 *The Manhattan Voyage*

On October 1968, Humble Oil, a private company for the U.S. acting on behalf of "Exxon" announced the voyage of *Manhattan* through Canada's portion of the NWP to prove that an icebreaking bulk carrier was capable of year-round sailings

[6] Article III, Annex A. In 1962 the Convention was amended by enlarging the restricted zones and changing the Conference of Contracting Governments to the Convention of 1954, London, April 4–11, 1962.

[7] 33 U.S.C. §§ 431–437 (Supp. II 1965–1966).

[8] *Oil in Navigable Waters Act, 1955*, 3 and 4 Eliz. 2, c. 25.

[9] *Oil in Navigable Waters Act, 1963*, c. 28.

[10] Official webpage of IMO, available at; http://www.imo.org/about/conventions/listofconventions/pages/international-convention-relating-to-intervention-on-the-high-seas-in-cases-of-oil-pollution-casualties.aspx (date accessed 27 February 2012).

[11] Adoption: 29 November 1969; Entry into force: 19 June 1975; Being replaced by 1992 Protocol.

[12] Adoption: 18 December 1971; Entry into force: 16 October 1978; superseded by 1992 Protocol.

between Alaska and the east coast of the United States.[13] There seemed to exist an intricate balance of threats and opportunities for Canada which established "jurisdictional" alongside "preservation of marine-environment" concern. Initially the government of the U.S. seemed co-operative, but steered clear of offering *de facto* support for Canada's "jurisdictional" claims.[14] American officials took the position that the *Manhattan* had navigated through high seas in the Passage because it had not traversed Canadian territorial waters, which at that time were distinctively embedded in the *Territorial Sea and Fishing Zones Act*[15] as extending for three miles from the islands of the Arctic Archipelago.[16] But Ottawa's intention to facilitate the *Manhattan* experiment followed with the motive to ensure *de facto* Canadian sovereignty and to avoid or delay a major confrontation over sovereignty.[17] The technical avoidance shifted gear from the spring of 1969 via public signals of determination on its position on sovereignty issues and a stronger emphasis on "jurisdictional" sovereignty dawned soon after.[18] The then Soviet Union tacitly signaled its support for Canada's efforts to enhance Coastal State control over Arctic waters. Nevertheless, *Manhattan's* voyage reflected the need for a definitive legal statement. It seems that two competing tendencies were evolving (in Canada) where the former one reflected the traditional approach to applying existing general International law to ice-covered waters (corresponding to International consensus) while the latter gave priority to pollution control over economic development by creating new International law in ice-covered areas.[19] "Consistent with International law" was, however, beginning to become less of a sacred principle as Canada had already asserted responsibility to itself on the use of this uniquely vulnerable environment and assumed to take the leading role in preserving its portion of this reserve.[20] This sense of urgency arose as the U.S.

---

[13] John Kirton and Don Munton, "The Manhattan Voyages and their Aftermath" in *Politics of the Northwest Passage*, Franklyn Griffiths (eds.), 1987, McGill-Queen's University Press, p. 70, where it is indicated that the *Manhattan* left an eastern seaboard port on 25 August 1969, to begin its Arctic journey, and broke through to ice-free waters off northers Alaska on 14 September. Eventually, it returned for a second voyage, confined to the area north of Baffin Island, in April–May 1970, *See* also *Supra* note 1 in Foreword.

[14] *Ibid.*, p. 71, para 2.

[15] *Territorial Sea and Fishing Zones Act*, S.C. 1964–1965, c. 22.

[16] *Supra* note 10, p. 72.

[17] *Ibid.*

[18] *Ibid.*, p. 73, reference may be drawn from the bill of 22 April introduced to amend the "Territorial Sea and Fishing Zones Act" to allow all waters above the Continental Shelf to be declared fishing zones exclusively for Canadian fishermen. This was duly criticized by the U.S. as contrary to International law, p. 76.

[19] *Ibid.*, p. 79.

[20] *Ibid.*, p. 81.

declared its second trip of *Manhattan*. The dawn of 1970 was marked with a firmly established policy of passing legislation to control shipping and pollution in the Arctic, exercising sovereignty through these actions.[21]

2.1.3 Legal Regime of AWPPA and ASPPR

Following the 1969–1970 *Manhattan* voyage, the Trudeau government enacted the AWPPA[22] creating a 100-mile environmental protection zone within Canadian Arctic waters.[23] Transport Canada has two legislative platforms for controlling the activities of ships operating in Canadian Arctic waters: CSA and the AWPPA. Under the dominion of Canada's Northern Strategy and in light of the assumed pioneering role, Canada has recently authorized the prolongation of the spatial scope of AWPPA from 100 to 200 nautical mile limit with complementary extension of the shipping safety control zones via Bill C-3.[24] The AWPPA and regulations adopted therewith, i.e. the ASPPR[25] are the original regulatory instruments to what comprises discharges from vessels in the Arctic. Moreover, the Canada Shipping Act 2001[26] and regulations incorporated therein, i.e. the Migratory Birds Convention

---

[21] *Ibid.*, p. 86.

[22] *Arctic Water Pollution Prevention Act* 1970 [R.S.C. 1985 (1st Supp.) C.2, (1st Supp.) S.1.], available at; http://www.tc.gc.ca/media/documents/acts-regulations/A-12-acts.pdf (date accessed 29 February 2012).

[23] *Supra* note 10, p. 67 where it is highlighted that on 8 April 1970, the government introduced into the House of Commons the Arctic Waters Pollution Prevention bill which was designed to prevent the pollution of waters adjacent to the mainland and islands of the Canadian Arctic, it asserted off-shore jurisdiction within a 100-mile pollution prevention zone.

[24] *Supra* note 49 in Chap. 1, *Note* that the extended definition of arctic waters is consistent with pre-existing provisions in the *Oceans Act*, which create Canada's exclusive economic zone, as well as the definition of the "sea" used in the *Canadian Environmental Protection Act, 1999*, which includes "any exclusive economic zone that may be created by Canada" for the purposes of provisions relating to disposal at sea. Moreover, it also provides an extended area of waters with respect to which the Governor in Council may establish Vessel Traffic Services Zones (VTS Zones) under Section 136 of the *Canada Shipping Act, 2001*.

[25] *Arctic Shipping Pollution Prevention Regulations*, C.R.C. Ch. 353, available at; http://laws.justice.gc.ca/PDF/Regulation/C/C.R.C.,_c._353.pdf (date accessed 29 February 2012), *Note* that the AWPPR applies to the deposit of waste in Arctic waters; or in any location on the mainland or islands of the Canadian Arctic; and the liability for such deposits. The AWPPR regulates the deposit of domestic and industrial waste in Arctic waters and on land in the Arctic, and the deposit of waste by ships in Arctic water. This Act also describes the limits of liability.

[26] *Canada Shipping Act, 2001* S.C., Ch. 26, available at; http://www.tc.gc.ca/media/documents/acts-regulations/C-10.15-acts.pdf (date accessed 29 February 2012), *Note* that *Canada Shipping Act* and associated Regulations such as the Regulations for the Prevention of Pollution from Ships and for Dangerous Chemicals (SOR/2007-86) constitute the main instrument regulating overall marine transportation in Canada including vessel-source pollution. Through these instruments Canada further implements MARPOL 73/78 Annexes II (noxious liquid substances), IV

Act 1994[27] and the Fisheries Act 1985[28] also accommodates provisions to what concerns vessel-source pollution.

In a hackneyed description of AWPPA, it may be stated that it is a "zero-discharge" Act, which states, "no person or ship shall deposit or permit the deposit of waste of any type in the Arctic waters" (Section 4 of AWPPA).[29] Tracing the law and to what concerns enforcement powers, The AWPPA describes offences and punishments; and outlines the powers that may be given to Pollution Prevention Officers so that they may enforce the Act (Sections 14 and 15 of AWPPA).[30] In this context it is germane to mention that this Act is a unique Act which acts with flexibility. The underlying reason for this comment is that with the growing trend to criminalize seafarers, examples of which are set forth by the European Union Directive 2005/35/EC as Amended by Directive 2009/123/EC, the AWPPA[31] only provides for monetary compensation even if the act was committed with *mens rea*.[32] This is also evident from Section 23(1) (a) which empowers the pollution

Footnote 26 (continued)
(sewage), V (garbage) and VI (air emissions). However, the discharge provisions provided for in the Pollution Prevention Regulation Sections 40 (oily mixtures), 82, 83 and 108 (noxious liquids and pollutants), 128 (sewage and sewage sludge), 139 (garbage) are not applicable to arctic waters *viz* ice-covered areas as safety control zones are expressly excluded.

[27] *Migratory Birds Convention Act*, S.C., 1994, Ch. 22, available at; http://laws.justice.gc.ca/PDF/Statute/M/M-7.01.pdf (date accessed 29 February 2012).

[28] Fisheries Act, R.S., 1985, C.F-14, available at; http://laws.justice.gc.ca/PDF/Statute/F/F-14.pdf. (date accessed 29 February 2012).

[29] *Supra* note 76 in Chap. 1.

[30] *Ibid.*

[31] Sections 18 and 19 of AWPPA which embodies "offences and punishment" in contravention to Section 4 and in the event of such contravention shall be subject to monetary compensation.

[32] *See DIRECTIVE 2005/35/EC of The European Parliament and of the council of 7 September 2005*; on ship-source pollution and on the introduction of penalties for infringements, available at; http://cleanseanet.emsa.europa.eu/docs/public/Directive_2005_35_EC.pdf (date accessed 29 February 2012), *See* also, *Europa* press release available at; http://europa.eu/rapid/pressReleasesAction.do?reference=MEMO/05/437&format=HTML&aged=1&language=EN&guiLanguage=en (date accessed 29 February 2012), where it is stated that the European Commission has used a recent judgment of the European Court of Justice (ECJ) to take powers to create new criminal penalties to enforce the entire body of European Union law. The ECJ Judgment (C-176/03), on 13 September, gave the EU jurisdiction to order prosecutions in cases of breaches of EU environmental law and on 23 November the Commission adopted a communication setting out its interpretation of the judgment of 13 September 2005 by which the Court of Justice annulled a framework decision on the protection of the environment through criminal law. As requested by the Commission, which had the backing of Parliament in its action against the Council, which for its part was supported by 11 Member States, the Court held that the Community had sole competence to take the criminal law measures needed to ensure the effectiveness of Community law. The judgment confers on the European Parliament a decisive role in the adoption of such measures whenever they are taken in a field governed by co-decision. This strengthening of parliamentary control, especially in areas as sensitive as criminal law was considered is a step forward for democracy.

prevention officers to only seize a vessel and its cargo whenever there are suspicions or reasonable grounds that any provision of AWPPA or ASPPR regulations adopted there under have been breached. The controlling regulation made under the AWPPA is the ASPPR which provides for construction and machinery standards for various classes of ice-strengthened vessels and establishes when and where in the Canadian Arctic such vessels can navigate based on their class and ice conditions in the area.

ASPPR embodies, pursuant to Section 4 of AWPPA, special permissions to what concerns only sewage and oil discharges which means for instance that the disposal of garbage or other noxious substances in Arctic waters is not allowed. Owing to the restrictive character of Canadian norms, Section 28 of ASPPR does not restrain the discharge of sewage generated on board ships in Arctic waters while Section 29(c) permits the discharge of oil only under very restricted circumstances when related to the safety of lives, or the vessel and to the normal operation of the engine or its components insofar as such discharges are minimal and unavoidable.[33] Section 17(2) of ASPPR corresponding to enforcement jurisdiction allows the annulment of a ships Arctic pollution certificate if upon inspection it is realized that there exists a probability that the ship is in danger of discharging or actually discharging waste into Arctic waters in violation of Section 4(1) of AWPPA.[34]

2.1.4 Absolute Liability and Strict Liability

The AWPPA as was introduced in the House of Commons on 8 April 1970, did not directly assert sovereignty over Canadian waters. It was an instrument to establish a pollution control zone and an extension of which was implemented via Bill C-3. The incorporation of such Act reflected the dire need to determine a jurisdiction over the function of the entire area, the jurisdiction of which was intended to be exercised preventively by setting shipping standards for that area and remedially by constituting a domestic liability and compensation regime.[35] To enforce the regulations, the legislation provided sanctions for pollution prevention officers with a broad enforcement powers. This coupled with the prohibition policy of "waste" in the Arctic waters leading to monetary penalty amalgamates the notion of "absolute liability", that is, like "strict liability" there was no need to prove that the damage was a result of negligence.[36] However, unlike strict liability, there

[33] *Supra* note 10.

[34] *Ibid.*

[35] R. Michael M'Gonigle and Mark W. Zacher, "Canadian Foreign Policy and the Control of Marine Pollution" in *Canadian Foreign Policy and the Law of the Sea*, Barbara Johnson and Mark W. Zazher (eds.), 1977, The University of British Columbia, p. 117.

[36] *Ibid.*

were no defenses available to a polluter. The Act does not specify the limits of liability, leaving it to be determined by the Act itself. This is the "absolute liability" regime which exists in Canadian waters North of 60 degrees. The liability facet of ship source oil pollution as is understood, is covered by CLC and amended by the CLC 1992, the elementary features of which is 'strict liability' of the tanker owner who is obliged to pay for damage resulted from 'persistent spill'. A significant compromise reached in the event of drafting the CLC is the notion of 'channelling of liability' where the liability for oil pollution would be channelled through to the 'registered owner' who is burdened with compulsory insurance.[37] Canada's initial impression on CLC being "insufficient" emanates from the idea where it devises a liberty for other parties and although recourse against those other parties may be initiated by the "registered owner", this had to be done after satisfying the claimant and sometimes it did not suffice to cover the entire pollution damage.[38] This alongside the massive pollution caused by the *Torrey Canyon* confirmed to Canada the necessity to draw the control line by the regime of "absolute liability". This status quo is, as the critiques would believe to be completely uneconomic and impractical for commercial interests since "absolute liability" exists as a potential removal of liability limitations and the power to unilaterally to set higher construction and operating standards.[39] As for "strict liability", Part XX of CSA integrates a comprehensive national regime for oil spill response, liability and compensation which was an innovative movement in the domain of domestic oil spill legislation.[40] This regime was endorsed by Canada prior to the International implementation and coming into force of CLC and FUND. CSA shrouded in antiquity, is one of the oldest pieces of legislation based on the *British Merchant Shipping Act*, 1894. It was amended over the years to suit contemporary purpose and in 1993, amendments to the CSA led to the creation of a network of private sector oil spill response organizations. Funded and operated by the private sector, the regime was

[37] Marc A. Huybrechts, *Criminal Liability of Master and Crew in Oil Pollution Cases*, Maritime Pollution Liability and Policy: China, Europe and the U.S., Edited by Michael G. Faure, Han Lixin & Shan Hongjun, 2010, Kluwer Law International, p. 216.

[38] The CLC and the Fund Convention as amended by their respective 1976 and 1992 Protocols (i.e. the CLC Convention 1992 and the Fund Convention 1992) now have the force of law in Canada pursuant to the Marine Liability Act, S.C. 2001, c. 6, Part 6 (Liability and Compensation for Pollution) (Sections 47–105) (in force 8 August 2001). Part 6 includes divisions on "Civil Liability for Pollution" (Sections 51–71) and on "Compensation for Pollution" (Sections 72–105). Part 8 (Sections 165–184) of the Canada Shipping Act, 2001, S.C. 2001, c. 26, deals with "Pollution Prevention and Response—Department of Fisheries and Oceans", while Part 9 (Sections 185–193) concerns "Pollution—Department of Transport", thus dividing the responsibility for the prevention of marine pollution in Canada between the Department of Fisheries and Oceans and the Department of Transport.

[39] *Supra* note 37.

[40] *Supra* note 26, which establishes the strict liability of owners to be responsible for costs and damages for a discharge of oil from all classes of ships.

established in 1995 to enable industry to respond to oil spills of up to 10,000 tonnes in Canadian waters south of 60 degrees north latitude.[41]

2.2 The Russian Federation and the NSR

2.2.1 Background

The NSR is synonymous to the Russian name for what is often known "outside Russia" as the Northeast Passage.[42] NSR activity was at its peak in 1987, but as the Soviet system started to crumble a perplexing situation arose for the state to uphold the high level of subsidies that was required to maintain most activities in the Arctic, and NSR cargo volumes diminished.[43] Gorbachev's *perestroika policy* was stamped with an alteration as regards geo-politics and economic objectives at the end of the Cold War, the change which supplemented receptive conditions for the opening of navigation in the NSR to foreign vessels.[44] Ever since the wake of that change, the Russian Federation has made efforts to confederate its concern and sovereignty as regards development and policy simultaneously in the Arctic.[45] In Russia, the term NSR holds different connotations, and evokes discernment of a grand national transport corridor significantly utilized for bringing natural resources out, and for bringing deliveries into the many settlements in the Russian Arctic.[46]

41 *Environmental Response*, Report of the Canadian Coast Guard Maritime Services, Ottawa, Ontario, 2009, Fisheries and Oceans Canada Publishers, available at; www.ccg-gcc.gc.ca/eng/Ccg/er_Home (date accessed 2 March 2012), where it is stated that all oil tankers of 150 tonnes gross tonnage and all other vessels of 400 tonnes gross tonnage trading in Canadian waters, as well as oil handling facilities located within Canadian jurisdiction, must have an arrangement with a response organization.

42 Claes Lykke Ragner, "Den norra sjövägen", In Hallberg, Torsten (ed.), *Barents—ett gränsland i Norden (Norden Association's Yearbook)*, Stockholm, Arena Norden, 2008, pp. 114–127, English translation available at; http://www.fni.no/doc&pdf/clr-norden-nsr-en.pdf (date accessed 3 March 2012).

43 *Ibid.*

44 Douglas R. Brubaker, "Regulation of Navigation and Vessel Source Pollution in the Northern Sea Route: Article 234 and State Practice" in *Protecting the Polar Marine Environment: Law and Policy for Pollution Prevention,* Davor Vidas (ed.) 2000, Cambridge University Press, p. 221, where it is indicated that this initiative eventually resulted in the formal opening of the NSR to non-Soviet vessels on 1 July 1991, only a few months before the Soviet Union was dissolved.

45 Aleksandr A. Kovalev and William Butler, *Contemporary Issues of the Law of the Sea: Modern Russian Approaches,* 2004, Eleven International Publishing, p. 180.

46 *Supra* note 42.

2.2.2 Jurisdiction and Regulations

Russia today claims formal jurisdiction over the NSR, based on Article 234 of UNCLOS. The Russian Regulations set out in the Guide to Navigation through the Northern Sea Route[47] incorporates that all vessels wishing to enter the NSR (*including* all areas within Russian 200 nautical miles exclusive economic zone) should give notifications to the Russian authorities beforehand. They must also submit an application for guidance and subject to paying a set fee to use the route, sporadically referred to as the "ice-breaker fee". Russia also claims the straits within and between the Russian Arctic archipelagos and the mainland as part of its internal waters. Other countries to a certain extent have accepted Russia's *de facto* control of these waters, and have not challenged the regime Russia has put in place. The U.S. with vigorous objections maintains that the straits should be considered international strait, and thus open to transit passage.[48] The Federal Law on the Internal and Territorial Marine Waters, Territorial Sea and the Adjacent Zone[49] establishes the basic provisions for the prevention of pollution of marine environment, where Article 37 stipulates that the discharges from vessels of harmful substances in these areas are prohibited and determines that operational discharges from vessels cannot exceed the permissible concentrations supplemented in other federal legislation. Moreover, The Federal Law on the Exclusive Economic Zone of the Russian Federation[50] is drafted with a similar structure, the concept of which is reflected in Article 4.[51] The sovereign rights of the Russian Federation in the EEZ is embedded in Article 5. The significant provisions of this Federal law are assumed to be Articles 32 and 33, which relates to vessel-source pollution in

[47] Guide to Navigating through the Northern Sea Route, Head Department of Navigation and Oceanography of the Ministry of Defense of the Russian Federation, Notice to Mariners 81–84 (13 July 1996), *See* also Erik Franckx, "The Legal Regime of Navigation in the Russian Arctic", *Journal of Transnational Law and Policy*, 2009, Vol. 18, no. 2, p. 337.

[48] *Supra* note 42.

[49] Federal Law of 31 July 1998, available at; http://www.un.org/Depts/los/LEGISLATIONANDTREATIES/PDFFILES/RUS_1998_Act_TS.pdf (date accessed 3 March 2012).

[50] Federal Law of 17 December 1998, available at; http://www.un.org/Depts/los/LEGISLATIONANDTREATIES/PDFFILES/RUS_1998_Act_EZ.pdf (date accessed 3 March 2012).

[51] Paragraph 13 of Article 4 of the Federal Law of 17 December 1998 states that, "Discharge of harmful substances or effluents containing such substances…: any discharge from vessels and other floating craft (hereinafter referred to as "vessels"), aircraft, artificial islands, installations and structures for any reason, …; discharge of harmful substances does not include the ejection of harmful substances occurring directly as a result of the exploration, exploitation and related treatment at sea of mineral resources of the continental shelf of the Russian Federation, or the discharge of harmful substances in order to conduct legitimate scientific research for the purpose of combating or monitoring pollution".

Arctic ice-covered areas. Observance with regard to the wordings of Article 32[52] would reveal its resemblance to Article 234 of UNCLOS which takes into cognizance special environmental characteristics of the Russian Arctic. Domestic Regulations[53] and Resolutions[54] with regard to operational discharge is parallel to MARPOL 73/74, but contains more stringent policies. Within the list of policies and regulations, it is significant to note the Law of the Russian Federation on Environmental Protection[55] concerning environmental regulations on vessel-source pollution in Arctic ice-covered areas and the Water Code of the Russian Federation[56] which invokes enforcement jurisdiction in the event of water contamination. Till date there exists various port regulations and federal prescriptions relating to NSR navigation[57] and the legal framework of it all which is in line with Maritime Doctrine and State Principles for the Arctic,[58] is undergoing substantive alterations to respond to the dawning demands of the Arctic waters.

[52] Article 32 of the Federal Law of 17 December 1998 states that, "With regard to areas which are within the limits of the exclusive economic zone and where particularly severe climatic conditions and the presence of ice covering such areas for most of the year … and pollution of the marine environment could cause major harm to or irreversible disturbance of the ecological balance, the Russian Federation may adopt and enforce federal laws and other regulations for the prevention, reduction and control of marine pollution. Such federal laws and other regulations shall have due regard to navigation and the protection and preservation of the marine environment and the natural resources of the exclusive economic zone based on the best available scientific evidence. The limits of such areas shall be published in *Notices to Mariners*." [Emphasis added].

[53] Regulations for Preventing the Pollution of Offshore Waters, Sanitary Regulations and Norms Preventing the Pollution of Offshore Waters in Water Supply Areas. They prescribe that the discharge of oil must comply with the requirements of MARPOL 73/78 for special areas.

[54] Russian Federation Government Resolution N° 251 of 24 March 2000 approving the list of denied toxic substances into the EEZ from ships and other floating equipment, aircraft, artificial installations or structures and Russian Federation Government Resolution N° 208 of 10 March 2000 establishing rules for the development and approval of norms of maximum permissible concentrations of harmful substances and permissible impacts on the marine environment and natural resources of internal waters and territorial sea.

[55] Federal Law on Environmental Protection of 10 January 2002, N 7-FZ, available at; http://www.icfinternational.ru/doc_files/oops.pdf (Russian version) (date accessed 3 March 2012).

[56] Water Code of the Russian Federation of 3 June 2006, N 74-FZ, available at; http://www.icfinternational.ru/doc_files/vodn_kodeks.pdf (Russian version) (date accessed 3 March 2012).

[57] *Supra* note 35 in Chap. 1, legislations which include the 1973 USSR Statute on State Maritime Pilots, the 1984 Edict on Intensifying Nature Protection in Areas of the Far North and Marine Areas Adjacent to the Northern Coast of the USSR, the 1985 Statute on the Protection of the Economic Zone of the USSR, the 1985 Statute on the Protection and Preservation of the Marine Environment in the Economic Zone of the USSR, Requirements for the Design, Equipment and Supply of Vessels Navigating the Northern Sea Route, Law on the Russian Federation's Internal Sea Waters.

[58] Maritime Doctrine of Russian Federation 2020, approved by President Vladimir Putin on 27.06.2001, available at; http://www.oceanlaw.org/downloads/arctic/Russian_Maritime_Policy_2020.pdf (date accessed 3 March 2012).

2.2.3 *Comparisons with AWPPA*

NSR is considered to be a nexus between the Russian Federation and the Arctic. Hence, development of law and policy to protect NSR is, in theory, protection of the Russian Arctic. In practice, protection of NSR from ship-source pollution is embodied in disparate random regulations. Although the federal laws give reference to "ice-covered areas", there still exists the need for a single predominant regulation which concentrates on the Arctic. The Law of the Russian Federation on Environmental Protection[59] although applicable to NSR is too general in a sense it does not give due regard to the current conditions of the Arctic. Examples may be drawn from AWPPA which in its "short title" provides a rather forthright insight into the Arctic waters by indicating "the internal waters of Canada and the waters of the territorial sea of Canada and the exclusive economic zone of Canada". Then again, the Water Code of the Russian Federation[60] provides for the designation of "Specially Protected Water Bodies" (Article 66) with a liability facet without including special conditions of the Arctic water. This Code, however, does not provide any definition of waters that are of specific relevance to the Arctic. The actual effectiveness of Russia's environmental legislation for the Arctic is said to be weakened by administrative, financial and operational constraints. This guarantees that principle policy statements corresponding to environmental protection rarely extend beyond official rhetoric.[61] On the other hand, para 3(2) of AWPPA narrows down the application of the Act to Arctic, while 3.1(1) supplements the option for amendments and enforcement of those amendments. Although Edgar Gold has reflected on the positive side of various national port regulations and federal prescriptions for the NSR,[62] it can be clearly deduced that the Russian Federation with the foregoing optimistic claim over NSR should amalgamate the provisions into a single regulation with relevant guidelines. AWPPA is, in fact, a model which might be taken into consideration to change the *status quo* of the Russian Arctic regime.

2.3 Arctic and the American Perspective

2.3.1 *Background*

During the height of the Cold War, the Arctic region was acknowledged as a geo-political and geo-strategic playground for the U.S. and the then Soviet Union.[63]

---

[59] *Supra* note 55.

[60] *Supra* note 56.

[61] J.D. Oldfield, "Russian Environmentalism", *European Environment: the Journal of European Environmental Policy*, 14 March 2002, doi: 10.1002/eet.286, Vol. 12, No. 2, p. 126.

[62] *Infra* note 62.

[63] Heather Conley and Jamey Kraut, *U.S. Strategic Interests in the Arctic: An Assessment of Current Challenges and New Opportunities for Co-operation*, "A Report of the CSIS Europe Program", Centre for Strategic and International Studies, April 2010, p. 1.

The U.S. has stood strong against Canada by pressing equally hard provisions guaranteeing "non-suspendable" transit rights in the Articles of UNCLOS pertaining to international straits.[64] Moreover, it has raised doubts as to the future status of the Convention with regard to its relations with Canada. The contemporary policy towards the Arctic was introduced in National Security Decisions Memorandum from 1971, which explicitly stated its concern in the rational development of the Arctic manoeuvered by the rationale of curtailing all adverse effects to the environment coupled with protecting the principle of freedom of the seas and superjacent airspace.[65] The National legislation emanates from different themes[66] and the end of the Cold War narrowed the concern of military security while heightening the other themes.[67]

2.3.2 *Acts for Arctic*

Relevant legislation relating to the Arctic is the Coastal Zone Management Act of 1972,[68] the participation of which is voluntary and where Alaska is one of the participating States. Federal assistance is afforded to any Coastal State to amplify and implement this comprehensive set of management infrastructure, thus encouraging states to participate in the Act itself.[69] The Oil Pollution Act of 1990[70] was adopted as a direct response to the *Exxon Valdez* incident. It incorporates "Prince William Sound Provisions" in Title V (Section 5001), which calls for the establishment of "Oil Spill Recovery Institute" where the significant function is to "identify and develop" convenient techniques, equipment and materials for "dealing with oil spills in the Arctic and sub-Arctic marine environment".[71] The legal bodies created under this Title are the "Advisory board", "Scientific and Technical Committee", "Director" and "Regional Citizen's Advisory Councils". Compliments may be rendered in view of its efforts to organize "Oil Terminal Facilities and Oil Tanker Operations Association", which is unique in the sense that it incorporates a balance between the categories of members, i.e. government, owners and operators of oil

---

64 *Supra* note 5 in Chap. 1, UNCLOS Articles 34–45.

65 Donald R. Rothwell and Christopher C. Joyner, "Domestic perspectives and regulations in protecting the polar marine environment: Australia, Canada and the United States" in *Protecting the Polar Marine Environment*, Davor Vidas (ed.), 2000, Cambridge University Press, p. 150.

66 *Ibid.*, i.e. military security; scientific security; economic security; and environmental security.

67 *Ibid.*, p. 157.

68 *The Coastal Zone Management Act*, 1972, as amended through Pub. L. No. 109-58, the Energy Policy Act of 2005, available at; http://coastalmanagement.noaa.gov/about/media/CZMA_10_11_06.pdf (date accessed 5 March 2012).

69 *Infra* note 59, p. 160.

70 *The Oil Pollution Act of 1990*, 33 United States Codes § 2701–2761 (1990), available at; http://epw.senate.gov/opa90.pdf (date accessed 5 March 2012).

71 *Ibid.*

tankers, owners and operators of terminal facilities and the locals of Alaska with a unique voting system. Then again, a significant comparison that may be drawn with Canadian AWPPA is that although it is a direct comprehensive Act dealing with pollution prevention in the Arctic, it only creates one entity, i.e. Pollution Prevention Officer with enforcement authorities. This is as opposed to Oil Pollution Act, 1990, which immaculately provides a comprehensive strategy coupled with an administrative system to prevent and control any major oil spills in the near future. Other principle Acts[72] are in operation under the Federal system among which Act to Prevent Pollution from Ships[73] is the U.S. enactment of MARPOL 73/78. Moreover, in the Arctic state of Alaska, operation and response are regulated by the Alaska Oil and Hazardous Substances Pollution Control Act and the Alaska Environmental Conservation Act.[74]

2.4 Other Legislative Efforts

Legislation in the Danish part of the Arctic is formulated via a complex interplay among various regulation levels, i.e. international, regional, national and local. This is to a greater extent impacted by alternative transnational regulatory prerequisites and regulation regimes. The activities in the Danish part of the Arctic is not governed by any specific legislation and the reason for avoiding specific conflict with its Arctic counterparts is due to the fact that the Kingdom of Denmark is centrally located in the Arctic.[75] The three parts of the realm, i.e. Denmark, Greenland and the Faroe Islands, share a number of values and interests and all have a responsibility in and for the Arctic region.[76] The Arctic makes up an essential part of the common cultural heritage, and is home to part of the Kingdom's population. The Kingdom of Denmark has developed this Arctic strategy (2011–2020) and does not intend to alter the power-sharing that exists between Denmark, the Faroe

[72] The 1980 Comprehensive Environmental Response, Compensation and Liability Act (42 United States Codes §§ 9601–9657 (1980)), the Federal Water Pollution Control Act (United States Codes §§ 151–160 (1948) as amended), the Trans-Alaska Pipeline Authorization Act (43 United States Codes §§ 1651–1655 (1973)), the Port and Tanker Safety Act (33 United States Codes §§ 1221–1236 (1978)), the Refuse Act (33 United States Codes §§ 407), the Marine Protection, Research and Sanctuaries Act (16 United States Codes c.32 (1972)).

[73] *The Act to Prevent Pollution from* Ships, 33 United States Codes §§ 1901–1903, available at; http://epw.senate.gov/atppfs.pdf (date accessed 5 March 2012).

[74] Edgar Gold, *Gard Handbook on Protection of the Marine Environment*, 3rd Edition, 2006, Arendal: Assuranceforeningen Gard, p. 369.

[75] Denmark, Greenland and Faroe Islands: Kingdom of Denmark Strategy for the Arctic 2011–2020, available at; https://web.law.columbia.edu/sites/default/files/microsites/climate-change/files/Arctic-Resources/Arctic-Council/02_01_Denmark%20strategy%20for%20the%20Arctic.pdf (date accessed 29.07.2014).

[76] *Ibid.*

Islands and Greenland, including responsibility for policy areas taken over and their respective funding.[77] Regardless of these individual distinctions, the Kingdom of Denmark has a common interest in addressing the challenges and utilising the opportunities arising from the Arctic region's rapidly changing conditions and growing interest from the world.[78] One of the areas Greenland has taken over is mineral resources. Decisions on development, exploration and exploitation of resources in Greenland are taken by the Greenland authorities. The Kingdom of Denmark has implicitly denoted that "a strategy" for the Arctic region is first and foremost a strategy for a development that benefits the inhabitants involving common interests relating to for example international agreements, and regional and global issues.[79] Then again, Denmark and Greenland's cooperation on Arctic indigenous peoples dates back to 1973 when the Arctic Peoples' Conference at Christiansborg Palace in Copenhagen became a launching point for the international organising of indigenous peoples.[80] Then again, *Denmark* has acknowledged the Greenlandic parliaments strong legislative authority, and Greenland thus retains extensive powers of self-government even though it remains under Danish rule. In addition, the Greenland Home Rule Government issued GHREO with the view to establishing a protected area around the Ilulissat Icefjord.

Studies from the U.S. Geological Survey, among others, estimate that there may be enormous, as yet, unproven oil and gas resources in the Arctic, just as previously major discoveries were made of gas especially (in Russia) but also oil (in Alaska). In Greenland's case, it is estimated that 31 billion barrels of oil and gas off the coast of Northeast Greenland and 17 billion barrels of oil and gas in areas west of Greenland and east of Canada could be discovered, though the probability is greater for discoveries in Northeast Greenland.[81] This is one of the significant reasons for the incorporation of the Arctic Strategy 2011–2020 where the vision is to exploit mineral resources in the Arctic under the best international practices, and in continued close cooperation with relevant authorities of the Danish Realm and international partners. It is also estimated that Greenland and the Faroe Islands shall be attractive areas for exploration, and the management of mineral resources must be competent and efficient in ensuring that such mineral resources are

[77] *Ibid.*

[78] *Ibid.*

[79] *Ibid.*

[80] *Ibid.*

[81] *Ibid.*, p. 24, where it has been stated that the mineral resources sector in Greenland has significantly matured over the last 10–15 years as a result of a long-term and deliberate strategy. After the adoption by Parliament Act No. 7 of 7 December 2009 on mineral resources and activities of relevance hereto, the mineral resources sector was fully taken over by the Greenland Self-Government on 1 January 2010 and is a key element to building growth industries and a self-sustaining economy. In 2008, Greenland had already adopted the Parliament Act No. 6 of 5 December 2008 on Greenland's Mineral Resources Fund, which is inspired by the Norwegian model so that oil and gas revenues also benefit future generations.

explored and exploited under the highest standards of environment and emergency preparedness and transparency.[82] Greenlandic legislation essentially applies GAIRAS in terms of shipping. Two regulations address ice navigation within Greenland's territorial sea: one specifies a number of rules of navigation while the other requires the fitting and use of ice searchlights.

Prior to formulation of the Arctic strategy, the DMA was founded in 1988 by merging the six Danish shipping authorities that existed at that time. The authority's mission is "to promote health and safety on clean seas and to effectively strengthen the competitiveness and employment of maritime industries". The DMA stipulates that all Danish vessels must be registered with the DMA and comply with their regulations and as such, it has jurisdiction over Danish vessels as well as vessels in Danish territorial waters.[83] The DMA plays an important

---

[82] *Ibid.*, where the strategy has given a strong focus on protection of the Arctic environment by stating that the mineral resource industry must be developed while strongly taking into consideration the fragile Arctic environment so it contributes to sound economic development, including the creation of new jobs and a maximum return to society. The strategy further provides that the mineral resource activities will also be carried out with sufficient preparedness that the public is kept from harm (based on the polluter-pays-principle) and that the Kingdom of Denmark's international obligations can be met in case of major unscheduled incidents. This should be a model for resource exploitation across the whole of the Arctic. This is followed by the fact that in the oil and gas sector, licensing rounds have been held biannually since 2002 and, alongside rising oil prices in recent years, there has been a breakthrough in the international interest in Greenland's oil potential. An area of more than 200,000 km. offshore South and West Greenland is now covered by 20 exploration and exploitation licenses, and in 2010 seven new exploration licenses were issued in Baffin Bay off the coast of Northwest Greenland. A licensing round for oil exploration off the coast of the northernmost part of East Greenland was held in 2012/2013. In the coming years in particular, there will be a need to maintain the level of activity offshore of West and Northwest Greenland, while ensuring a broad professional knowledge building in the more inaccessible areas off Northeast Greenland. With respect to oil and gas finds and subsequent production, a number of new challenges and tasks will emerge. As a result, coordination and cooperation with neighbouring Arctic states with similar challenges will be a major priority. Before new offshore areas are designated as oil/gas licensing areas, the Government of Greenland sets in motion the preparation of strategic environmental impact assessments in order to ensure that any oil/gas activities can be implemented on an environmentally sustainable basis. The strategic environmental impact assessments are prepared on a scientific basis by the National Environmental Research Institute and Greenland's Institute of Natural Resources. In connection with an application for the carrying out of concrete oil/gas activities which are likely to have a significant impact on the environment, such as exploration wells, the licensee is required to conduct a specific assessment of the environmental impact. The EIA report is submitted for public hearing and must be approved by the Government of Greenland before the application to carry out the activity can be approved. Under the Mineral Resources Act, companies seeking a license for exploitation must also prepare an Assessment of Societal Sustainability. The report must, for example, describe the utilization of Greenland manpower and enterprises in the project, including how the proportion of Greenland employees and subcontractors can be increased through training and skills development.

[83] Official Website of the Danish Maritime Authority, available at; http://www.dma.dk/Policy/Sider/Mainpage.aspx, (date accessed 29.07.2014).

role in the safety management aspect in Arctic navigation. Moreover, the ICES is the organization that coordinates and promotes marine research in the North Atlantic and has its headquarters in Denmark. Article 2 of its convention states that it is concerned with the Atlantic Ocean but this includes adjacent seas such as the Baltic Sea and North Sea. ICES act as a meeting point for scientists from the 19 member countries around the North Atlantic as well as other affiliates countries around the world (Australia, Chile, Greece, New Zealand, Peru and South Africa).

In 2007, the Icelandic Minister for Foreign Affairs declared in his annual speech to the Icelandic Parliament that the Arctic was a new core element in Iceland's foreign policy.[84] The 1979 Act Concerning the Territorial Sea, the Economic Zone and the Continental Shelf demarcates Iceland's maritime zones: all of the territorial sea baselines are drawn according to the straight baseline method. Although Iceland has been defined by the Arctic Council as belonging entirely to the Arctic region, Iceland's maritime zones do not form a part of the Arctic Waters as defined in the IMO's Polar Shipping Guidelines.[85] The 2004 Act on Marine and Coastal Anti-Pollution, as amended, constitutes the principal item of legislation concerning marine pollution by ships.[86] Applicable shipping standards do not go beyond GAIRAS which is similar to the concept of the Kingdom of Denmark and relevant EU standards. Iceland is a party to the EEA agreement as well as IMO. Iceland has not sought to rely on Article 234 of the UNCLOS to implement special arrangements in ice-covered maritime waters. From a critical analysis of the Act, it is clear that the "management and organization" segment of the Act is quite comprehensively marked out with detailed supervision and monitoring aspects.[87] The important part that deals with protection of the Arctic environment is embedded in Article 6 which is rather detailed in contents

[84] Legal Aspect of Arctic Shipping, *Legal and socio-economic studies in the field of the Integrated Maritime Policy for the European Union*, Summary Report of the European Commission, 2010, European Union. ISBN 978-92-79-15828-5, doi: 10.2771/51132.

[85] *Ibid.*

[86] The 2004 Act on Marine and Coastal Anti-Pollution, Iceland Act no. 33/2004 on marine and coastal antipollution measures, where Article 1 establishes the goals the Act, "[t]he goal of the 2004 Act on Marine and Coastal Anti-Pollution act is to protect the ocean and beaches of Iceland against pollution and activities that could jeopardize human health, damage the ocean's living resources and disturb its biosphere, damage the environment or prevent the lawful utilization of the ocean and beaches. It is also the goal of this act that after an accident resulting in pollution, the environment will be restored to its former state."

[87] *Ibid.*, where Article 4 stipulates that the Icelandic Coast Guard, under the supervision of the Minister of Justice, sees to monitoring the ocean areas around Iceland, from the air as well as on the sea. The Icelandic Coast Guard, if it spots pollution, or pollution of the ocean or coast is suspected, notifies the Environment and Food Agency and police of areas from which pollution can spread to the land. The Icelandic Maritime Administration, under the supervision of the Minister of Communications and Transport, is responsible for monitoring ships' pollution control equipment, cf. the Act on monitoring of ships.

compared to other national legislation corresponding to the Arctic.[88] Another significant effort is noteworthy in "Iceland's Arctic Policy"[89] which secures Iceland's position as a coastal State within the Arctic region as regards influencing its

---

[88] *Ibid. N.B.* Article 6 which states that the Minister for the Environment, after receiving proposals from the Environment and Food Agency, and in consultation, depending on relevance, with the Minister of Justice, Minister of Fisheries, Minister of Communications and Transport and the Union of Local Authorities in Iceland, sets general provisions into regulations concerning: (a) discharge of oil and oil mixed with water into the ocean from ships more than three nautical miles from the baseline of the territorial waters, including special ocean areas, as well as from platforms and other structures within Iceland's pollution jurisdiction beyond three nautical miles from the baseline of the territorial waters; (b) limitation of the quantity of oil in drainage water that may be channelled into the sea; (c) best available antipollution technology and the best environmental practices where such have been defined; (d) the equipment of ships, platforms and other structures in the open ocean and the equipment of companies on land for the prevention of marine and coastal pollution caused by oil and the monitoring of this equipment; (e) collection and destruction of waste oil, including the receipt of waste while from ships in harbours; (f) limitation or prohibition of this charge of fish oil and foots; (g) classification of liquid chemicals transported in ships' cargo holds to or from Icelandic harbours, as well as provisions on the limitation of discharge of chemicals deemed noxious, into the ocean beyond three nautical miles from the baseline of the territorial waters; (h) classification of substances used for antipollution measures under this act; (i) discharge of sewage into the ocean from ships, platforms and other structures on the open ocean; (j) handling and discharge of refuse from ships into ocean areas beyond three nautical miles from the baseline of the territorial waters; (k) receiving and handling of waste from ships; (l) receiving facilities for sewage from ships and other waste not listed above and its destruction; (m) limitation or prohibition of the discharge of ballast water from other sea areas, to prevent exotic organisms from being transmitted to Iceland; (n) prohibition or limitation of the discharge of substances into the sea from land that are listed in Annex II in this act; (o) throwing overboard of substances or objects into the ocean; (p) how the laying of submarine cables, submarine pipes or any kind of structure on the ocean floor shall be organized; (q) surveillance and measurements, such as to monitor possible changes in pollution of the ocean, and what research and measurements pollutants in the ocean and marine organisms and on the ocean floor, shall be made; (r) operating licences for business operations falling under this act; (s) transport of dangerous products by ship, where it is authorized to refer to the original foreign version of a list of substances and standards approved by the International Maritime Organisation; (t) preventive measures for and responses to acute pollution, operation of antipollution equipment, the duty to provide information to and cooperate with inspectors; (u) monitoring, recording and duty to notify; (v) prohibition or limitation of pollution from ships, platforms and other structures at sea or from land stations in accordance with the annexes of the MARPOL 73/78 protocol and other international agreements of which Iceland is a member; (w) guarantees and insurance because of business activities with objective liability for acute pollution in accordance with the current provisions of law; (x) transfer of substances, cf. Annex II, within Iceland's pollution jurisdiction.

[89] *A Parliamentary Resolution on Iceland's Arctic Policy*, Approved by Althingi at the 139th legislative session, March, 28 2011, available at; https://web.law.columbia.edu/sites/default/files/microsites/climate-change/files/Arctic-Resources/Arctic-Council/04_Iceland%20Arctic-Policy-Approved-by-Althingi.pdf (date accessed 29.07.2014), where the second principle implies that this will among other things be based on the fact that since the northern part of the Icelandic Exclusive Economic Zone falls within the Arctic and extends to the Greenland Sea adjoining the Arctic Ocean, Iceland has both territory and rights to sea areas north of the Arctic Circle. The Government shall in parallel develop the arguments which support this objective, in cooperation with relevant institutions.

development as well as international decisions on regional issues on the basis of legal, economic, ecological and geographical arguments. Then again, a significant development of the IAP is based on the notion that the Arctic Ocean should not be limited to a narrow geographical definition but rather be viewed as an extensive area when it comes to ecological, economic, political and security matters. Mutual cooperation as a part of the policy is established for strengthening and increasing cooperation with the Faroe Islands and Greenland with the aim of promoting the interests and political position of the three countries. This together with the Kingdom of Denmark's Arctic policy brings the four Arctic States on a common platform that seeks to promote and safeguard the Arctic marine environment. The IAP also recognizes the rights of the Arctic indigenous inhabitants by supporting their rights, which is operating in close cooperation with indigenous organizations and supporting their direct involvement in decisions on regional issues.[90] In addition, the CSR aspect has not been ignored in the IAP, which suggests the use of all available means to prevent human-induced climate change and its effects in order to improve the well-being of Arctic residents and their communities.[91] Iceland's NCA is also applicable to the terrestrial area, the territorial waters, the EEZ and the continental shelf.[92] The conservation and utilization of resources shall be

---

90 *Ibid.*

91 *Ibid.*, where it is stated that Iceland will concentrate its efforts fully on ensuring that increased economic activity in the Arctic region will contribute to sustainable utilization of resources and observe responsible handling of the fragile ecosystem and the conservation of biota. Furthermore, to contribute to the preservation of the unique culture and way of life of indigenous peoples which has developed in the Arctic region. But all this is subject to mutual assistance and cooperation among the Arctic State members whereby the IAP ensures that an understanding should be promoted of the fact that the This dues to the fact that the Arctic region both extends to the North Pole and the portion of the North-Atlantic region which has closest ties with the Arctic. The Arctic region should therefore be regarded as a single vast area in an ecological, political, economic and security-related sense, but not in a narrow geographical sense with the Arctic Circle, tree line or a temperature of 10 degrees centigrade in July as a reference point. Such a definition of Iceland's interests includes relations with other States, within the Nordic cooperation, defence cooperation with the United States, regional defence and security cooperation with Norway, Denmark and Canada, cooperation with the other seven Arctic States in the Arctic Council, relations with the European Union through participation in the so-called Northern Dimension (a cooperative forum including Russia, the EU, Iceland and Norway) and cooperation with Russia in the Arctic Council, through the Barents Euro-Arctic Council and within the Northern Dimension.

92 The Nature Conservation Act no. 44/1999, Ministry for the Environment of Iceland, available at http://eng.umhverfisraduneyti.is/legislation/nr/389 (date accessed 31.07.2014), lays down the objective of the Act which stipulates that "[t]he purpose of this Act is to direct the interaction of man with his environment so that it harm neither the biosphere nor the geosphere, nor pollute the air, sea or water. The Act is intended to ensure, to the extent possible, that Icelandic nature can develop according to its own laws and ensure conservation of its exceptional or historical aspects. The Act shall facilitate the nation's access to and knowledge of Icelandic nature and cultural heritage and encourage the conservation and utilisation of resources based on sustainable development." Compared to the other Arctic States, it seems that Iceland has proceeded to protect its share of the Arctic environment with necessary regulations and since Iceland is considered to

based on sustainable development which is considered a unique feature of the NCA in the context of marine protection of the Arctic. The NCA in this respect is a comprehensive Act dealing with rights and responsibilities coupled with compensation for damage to the environment.

FSAR, i.e. Finland's Strategy for the Arctic Region constitutes the goals of Finland's Arctic policy and means for their promotion.[93] It also explains the region's security, environment, economy, *natural resources,*[94]

Footnote 92 (continued)

belong completely to the Arctic zone, it seems that it has accomplished in implementing safeguard measures to prevent pollution in the sensitive waters of the Arctic. This is also evident from Article 13 which establishes that "[p]ersons travelling through the country must show full consideration for land owners and others with special rights to the land, respect their interests, for instance with regard to livestock or cultivation, including silviculture or reclamation land, and abide by their directions and instructions concerning travel and treatment of the land. Marked routes and paths and tracks provided shall be followed wherever possible, fences shall not be disturbed, by using gates or stiles wherever possible; when using gates care shall be taken to close them after passing through. Special caution shall be exercised in the vicinity of livestock, seal grounds, nesting areas, and hunt-ing and fishing areas. Passage through an area is not the responsibility of the land owner or rightholder in other respects than may result from the provisions of other Acts and general rules on damage compensation."

[93] *Finland's Strategy for the Arctic Region*, Prime Ministers Office Publications, 5 July 2010, ISBN: 978-952-5896-10-7, available at; http://web.law.columbia.edu/sites/default/files/microsites/climate-change/files/Resources/FinlandArctic.pdf (date accessed 30.09.2014). Finally, the "protection of sites of natural interests at sea" section is incorporated in Article 54 which directly states that "[t] he Minister for the Environment may, after receiving the approval of the Minister of Fisheries and, as the case may be, after receiving the proposals or opinion of the Marine Research Institute, the Nature Conservation Agency, the Icelandic Institute of Natural History or the Nature Conservation Council, protect sites of natural interest at sea within the territorial sea and the exclusive economic zone, including islands and skerries, and on the seabed, which it is deemed important to preserve because of their beauty or special characteristics or which it is important, from a scientific, natural historic or other cultural perspective, not to disturb. Areas surrounding sites of natural interest shall also be protected as necessary for them to be enjoyed to best advantage; such shall be clearly stated in the declaration of protection. The provisions of other Articles of this Chapter shall apply as appropriate to protected sites of natural interest at sea."

[94] *Ibid.*, *N.B.* that the oil and gas reserves in the Arctic Region play a key role for European energy supply. According to various estimates, 5–13 % of the world's untapped oil reserves and 20–30 % of gas reserves are located in the Arctic. As the old reserves in Russia gradually run dry, maintenance of the same level of gas and oil production requires considerable investments and the adoption of new technology. In fact, the emphasis in gas production is shifting from Western Siberia to deposits located in the continental shelf, in particular. The largest known deposits that will probably be exploited first are those in the northern parts of the Yamal Peninsula and the Shtokman and Fedinski (in Norwegian: Hjalmar Johansen-høyden) fields in the Barents Sea. It is expected that shrinking and thinning of the ice cover in the Arctic Ocean will facilitate the utilization of hydrocarbon reserves at sea. However, it should be noted that much of the hydrocarbon reserves in the Arctic are difficult to take into use, both economically and technically. In the future, the increased production of shale gas may weaken the competitiveness of Arctic gas. In addition to hydrocarbons, the mining industry offers much development potential both in Finland and in neighbouring

infrastructure,[95] indigenous people, international institutions and the European Union's Arctic policy. Although it moves to highlight multifarious issues, it seems that protection of the Arctic environment and climate change is given due importance among other environmental factors.[96] The fundamental observations of the Strategy deals with the utilization of Finland's Arctic *know-how,*[97] research, strengthening of the Arctic Council and development of the EU's Arctic policy.

Footnote 94 (continued)
countries. The Finnish mining industry has the opportunity to increase technology exports by networking, for instance, with foreign mining companies operating in Finland. In the future, the mining industry will need new technology for mining projects and a wide range of logistics investments in railways, roads, ports and handling equipment. One of Europe's biggest gold mines was opened in Kittilä, Finland in 2009, and other mines are being planned. Finnish mining technology is of a high standard in terms of environmental protection and technical aspects. The Arctic region has living natural resources that are diverse and abundant. A considerable percentage of the world's fish catch comes from the northern sea areas. It is difficult to assess the overall effects of climate change on the fish stocks of the Arctic Region. For Finland, it is important to guarantee the preservation and viability of natural fish stocks and other Arctic species in the Northern Sea areas and in Finland's northern rivers that empty into the Arctic Ocean. Preserving the viability of fish stocks requires efficient management of fishing and measures to prevent overfishing. Apart from fishing, the commercial utilization of plant and animal species (bioprospecting), for instance for making medicines, is on the increase. In Finland, reindeer husbandry is also a locally important rural trade, which is the source of livelihood for about a thousand families.

[95] *Ibid., N.B.* Finland's heavy dependence on foreign trade conducted mainly by means of sea transports sets challenges for the development of transport infrastructure and the logistic system. As the political and economic importance of the Barents Region increases, the need to develop the transport system, border crossings and telecommunications links will persist for a long time. The development of transport and logistics networks in Northern Finland also requires shared views with neighbouring countries. The melting of the ice in the northern sea routes may transform the world's logistics flows in the future. However, according to even the most optimistic assessment, safe year-round traffic in the Northeast and Northwest Passages will not be possible until some decades from now. In consequence, the Baltic Sea will remain the main channel for Finland's sea transports in the coming years. However, Finland must take note of the opening of the northern sea routes when strategic decisions about transports are made.

[96] *Ibid.*, where it is stated that the Copenhagen Accord reached at the United Nations Climate Change Conference in 2009 emphasizes the importance of reducing greenhouse gas emissions; which would help keep the increase in global temperature below 2 °C. At the same time, FSAR proceeds to state that provision needs to be made for the negative effects of climate change. The Finnish policy for adapting to climate change must also pay special attention to measures that would support the adaptation of livelihoods dependent on the Arctic environment. Another priority is the use and management of water resources, including the risks arising from more frequent floods.

[97] *Ibid.*, this is basically interpreted as Finland's objectives to strengthen, to make better and to improve expertise, experience and opportunities respectively. *Note* also, p. 18, where it is mentioned that, "[n]atural Natural resources and logistic connections offer opportunities in which Finland wants to invest. From the perspective of Finnish—especially Northern Finnish—industry and employment, it is important that all types of economic activity increase both in large seaports and in the land-based support areas of oil and gas fields in Norway and Russia. As economic activity increases, it is vital to ensure that operations in the entire Arctic Region are anchored in best practices and sustainable development, where also the status and rights of indigenous peoples are respected."

The Strategy was devised by a working group appointed by the Prime Minister's Office, for which all ministries were able to name their representatives. The Strategy accentuates external relations between Finland and its Arctic neighbours and was submitted to the Finnish Parliament as a report. The Advisory Board on Arctic Affairs, appointed by the Government, will play a central role in future work on the report. Finland has a diverse set of regulations some of which are applicable to the Arctic and this effort is innovative compared to legislation of other Arctic States. While other Arctic counterparts has set forth a detailed instrument in a single piece of document, Finland on the other hand has delved into disparate areas of national interest some of which are undertaken to fulfil EU and international obligations. For example, the organization of river basin management in Finland is covered by the AWRM (1299/2004),[98] the GDWRM (1040/2006)[99] and the related GDWRMR (1303/2004).[100] These legislations form part of Finland's ongoing implementation of the EU Water Framework Directive. Then again, legislation designed to prevent the pollution of water bodies is included in broader environmental protection statutes that came into force in the year 2000, which includes EPA (86/2000),[101] EPD (169/2000)[102] and DSDHAE (1022/2006).[103] Other legislation includes AEPMT (1672/2009)[104] and AWS (119/2001).[105] AEPMT was promulgated with the objective to prevent environmental pollution resulting from the normal operations of ships and gives reference to MARPOL 73/78, the AFS Convention and the TBT Regulation.[106]

---

98 Act on Water Resources Management 1299/2004, Given in Helsinki on 30 December 2004, available at http://www.finlex.fi/en/laki/kaannokset/2004/en20041299.pdf (date accessed 31.07.2014).

99 Government Decree on Water Resources Management (1040/2006) Given in Helsinki on 30 November, 2006, available at http://www.finlex.fi/en/laki/kaannokset/2006/en20061040.pdf (date accessed 31.07.2014).

100 Government Decree on Water Resources Management Regions (1303/2004), Issued in Helsinki 30 December 2004, available at; http://www.finlex.fi/en/laki/kaannokset/2004/en20041303.pdf (date accessed 31.07.2014).

101 Environmental Protection act, (86/2000; amendments up to 728/2011 included), February 4, 2000, available at; http://www.finlex.fi/en/laki/kaannokset/2000/en20000086.pdf (date accessed 31.07.2014).

102 Environmental Protection Decree, No. 169/2000, Issued in Helsinki 18 February 2000, available at; http://www.finlex.fi/en/laki/kaannokset/1999/en20000169.pdf (date accessed 31.07.2014).

103 Government Decree on Substances Dangerous and Harmful to the Aquatic Environment 1022/2006, Issued in Helsinki 23 November 2006, available at; http://www.finlex.fi/en/laki/kaannokset/2006/en20061022.pdf (date accessed 31.07.2014).

104 Act on Environmental Protection in Maritime Transport, (1672/2009), available at; http://www.finlex.fi/en/laki/kaannokset/2009/en20091672.pdf (date accessed 31.07.2014).

105 Water Services Act, (119/2001), available at http://www.finlex.fi/en/laki/kaannokset/2001/en20010119.pdf (date accessed 31.07.2014).

106 *Supra* note 104, where Section 1(1) of the Act incorporates the aims and objectives which states that "[t]he purpose of this Act is to prevent environmental pollution resulting from the normal operation of ships, by prohibiting discharges and emissions of noxious substances into water and air, or by setting limits on discharges and emissions into water and air. Furthermore, the purpose of

Sweden has implemented its own set of legislations in order to give effect to its obligations as a part of *pacta sunt servanda* pursuant to MARPOL 73/78. The Act relative to measures against pollution caused by ships contains prohibitions on the discharge of oil and oil, to that extent, may not be discharged in Swedish maritime areas; the maritime areas of another EU country; on the high seas; or in defined special areas.[107] From a more downright Arctic perspective, Sweden has given effect to its own SSAR, whereby the purpose of the Government's Strategy for the Arctic Region is to present Sweden's relationship with the Arctic, together with the current priorities and future outlook for Sweden's Arctic policy, proceeding from an international perspective.[108] The strategy begins with a summary, followed by an introduction of Sweden as a part of the Arctic conglomeration.[109] Further, it specifies how, and through which international cooperation bodies and bilateral channels, the Government should seek to secure and achieve its objectives for the Arctic.[110] Finally, it discusses the top priorities in the strategy's three

Footnote 106 (continued)

this Act is to organise the reception of waste in ports resulting from the normal operation of ships." Then again Section 1(2) highlights the fact that this Act lays down the provisions for the national implementation of international commitments binding on Finland and European Community legal instruments pertaining to the prevention of environmental pollution resulting from the normal operation of ships, as well as other provisions pertaining to the prevention of environmental pollution resulting from the normal operation of ships. Chapter 2 of the Act establishes the International scope in the application of the Act itself by implementing International standards stating that "[i]t is prohibited to discharge oil or oily mixtures from ships in Finnish waters or in Finland's exclusive economic zone, as well as from Finnish ships outside Finnish waters or Finland's exclusive economic zone, as set out in Annex I to the MARPOL 73/78 Convention, in the Helsinki Convention, or in other international commitments binding on Finland, or in European Community legal instruments." For a violation of the prohibition, laid down in Chapter 2, Section 1, on the discharge of oil or oily mixtures in Finland's territorial waters or Finland's exclusive economic zone, a monetary penalty (oil discharge fee) has been imposed in Chapter 3, Section 1. It adds the exception if the discharge is deemed minor in amount and impact. However, an oil discharge fee shall be imposed by the power of this Act on foreign ships in transit for any violation of the discharge prohibition in Finland's exclusive economic zone, only if the discharge causes considerable damage or risk of damage to Finland's shoreline or to the interests pertaining thereto, or to the natural resources in Finland's territorial sea or within Finland's exclusive economic zone.

107 Act Relative to measures against pollution caused by ships (Translated version) [Förordning (1980:789) om åtgärder mot förorening från fartyg], available at; http://www.notisum.se/rnp/sls/lag/19800789.htm (date accessed 31.07.2014).

108 Sweden's Strategy for the Arctic Region, Regeringskansliet, Government offices of Sweden, 2011, available at; https://web.law.columbia.edu/sites/default/files/microsites/climate-change/files/Arctic-Resources/Arctic-Council/Sweden%20Arctic%20Policy.pdf (date accessed 31.07.2014).

109 *Ibid.*

110 *Ibid.*, *N.B.* a most significant aspect of SSAR is to focus on the facet that Sweden will also strive to strengthen the Arctic Council in its role as the central multilateral forum for Arctic–related issues, as well as the role of Barents cooperation bodies in issues of particular relevance to the Barents region. The suggestion is that a more common policy and concrete projects should be developed in Arctic-related cooperation forums for the benefit of the region. Then again, SSAR points out that Sweden will actively contribute to the development of an EU Arctic policy

thematic areas: climate and the environment, economic development, and the human dimension.[111] This is noted as the first strategy the Government of Sweden has adopted on the Arctic as a whole, and should be seen as a starting-point for further development of cooperation in the region.[112]

In Norway, the development in the High North, including the Arctic has been on the government's agenda since 2005.[113] The overreaching aim was to achieve greater knowledge and to lay the foundations for sustainable economic and social development in terms of the sensitive marine environment.[114] As such, the HNS was established in 2006, which was followed by the report "New Building Blocks in the North" which identifies several priority areas.[115] Norway has also

---

Footnote 110 (continued)

and the fact that Sweden wishes to promote the EU as a relevant cooperation partner in the High North within relevant policy areas. Moreover, in the Nordic Council of Ministers, Sweden will work to sharpen the focus of Arctic-related project activities that have a clear supplementary value for the Arctic Council.

[111] *Ibid.*, *N.B.* this is where the SSAR proceeds to explain climate and environmental ties in relation to Arctic and Sweden. Sweden's climate and environment are a part of the Arctic and as a result both affect and are affected by it. One challenge for the Swedish Government will be to deal with the increase in precipitation caused by global warming, which may lead to greater water flows and changes in soil conditions. This in turn may affect Swedish societies and their infrastructure. The Sámi culture and industries traditionally have strong links to the surrounding natural environment and weather conditions, leaving them particularly vulnerable.

[112] *Ibid.*

[113] *Ibid.*

[114] *Ibid.*

[115] The Norwegian Government's Hight North Strategy, the Norwegian Ministry of Foreign Affairs, 2006, available at; http://www.regjeringen.no/upload/Ud/Vedlegg/strategien.pdf (date accessed 31.07.2014), where the strategy sets out the framework for the Norwegian efforts in the High North. In its policy platform, the Norwegian Government states that it considers the High North to be Norway's most important strategic priority area in the years ahead. Norwegian interests in the High North under the strategy will be safeguarded primarily by strengthening our presence and increasing the level of activity in a number of policy areas at both national and international level. These include education and research, environmental and resource management, safety and emergency response systems, energy, fisheries, tourism and other economic activities, health, culture and gender equality. Then again, the strategy acknowledges that an ecosystem-based management of resources requires continuous assessment of the state of the environment in relation to the environmental goals that have been set. The introduction of a coordinated monitoring system for the state of the marine environment, including pollution levels, will require further research to establish background levels for environmentally hazardous substances and continue the development of indicators. In this regard, the implementation of the MAREANO programme to develop a marine areal database for Norwegian waters will increase knowledge of physical, biological and chemical conditions on the seabed. MAREANO will be an important tool in establishing integrated ecosystem- based management regime and providing a framework for increased value creation. Then again, the Government commits to promote commercially and environmentally sustainable development of marine biotechnology in the High North. This will involve strengthening knowledge generation in the field, as well as increasing the technological, industrial and marketing expertise needed to realize the opportunities for value

implemented legislation for the protection of the environment in Svalbard.[116] The Act relating to the PES entered into force in 2001 whereby the objective of the Act is to preserve a "virtually untouched environment" in Svalbard. The Act is applicable to the entire land area of Svalbard and its waters out to the territorial limit. The Act incorporates the precautionary principle, "the polluter pays principle"[117] and the utilization of the best available techniques for activities that take place in Svalbard. It also stipulates the implementation of protected areas encompassing all habitats and landscape types, for the purposes of maintaining areas of historical value and to protect ecosystems on land and in the sea.[118] Then again, the environmental protection authorities for Svalbard include the King, the ministry, directorates as is decided by the ministry and the Governor of Svalbard.

Footnote 115 (continued)
creation offered by marine bioprospecting. Framework conditions and legislation are currently being developed to govern commercial and intellectual property rights to the genetic resources in the sea and on the seabed. Moreover, the "New Building Blocks in the North" identifies the following areas; (1) climate and the environment; (2) monitoring-emergency response-maritime safety in northern waters; (3) sustainable development of offshore petroleum and renewable marine resources; (4) onshore business development; (5) infrastructure; (6) sovereignty and cross-border cooperation; and (7) the culture and livelihoods of indigenous peoples.

[116] Act of 15 June 2001 No. 79 Relating to the Protection of the Environment in Svalbard, Amended by the Act of 20 April 2012 No. 20 (in force from 1 July 2012 in accordance with the Royal Decree of 20 April 2012 No. 332), available at; http://www.regjeringen.no/en/doc/laws/acts/svalbard-environmental-protection-act.html?id=173945 (date accessed 31.07.2014), where Section 1 enumerates that "[T]he purpose of this Act is to preserve a virtually untouched environment in Svalbard with respect to continuous areas of wilderness, landscape, flora, fauna and cultural heritage. Within this framework, the Act allows for environmentally sound settlement, research and commercial activities."

[117] *Ibid.*, where Section 9 stipulates that "[t]he cost of preventing or limiting damage to the environment or cultural heritage shall be covered by the person that is or would be the cause of such damage. Likewise, the cost of preventing or limiting pollution and waste problems shall be covered by the person that is or would be the cause of such problems."

[118] *Ibid.*, where Section 12 incorporates that "[t]he The regulations shall indicate the boundaries of the protected area and its purpose, and include provisions governing the use of the area. The King may in the regulations prohibit or regulate any activity and access or passage that, on its own or in combination with other types of use, is liable to undermine the objectives of such protection. This is in conjunction with Section 13 which establishes that "[t]he At an early stage in the preparation of regulations under Section 12 regarding the establishment or extension of a protected area or the repeal of its status, or regarding significant amendments to the protection provisions for a protected area, the Governor shall ensure cooperation with public authorities and organisations etc. that have a special interest in the decision. The Governor shall publish a notice in at least one newspaper with a wide circulation in Svalbard giving an account of the planned protection measures. Landowners and holders of rights shall as far as possible be informed in writing and given a reasonable time limit for expressing an opinion before the proposal for protection measures is drawn up. During the preparation of such proposals, the consequences for relevant activities in the area shall be clarified."

implemented legislation for the protection of the environment in Svalbard. The Act relating to the PES entered into force in 2002 [illegible] the Act is to preserve a "virtually untouched environment" in Svalbard. The Act is applicable to the entire land area of Svalbard and its waters out to the territorial limit. The Act incorporates the precautionary principle, [illegible] principle and the obligation to use the best available techniques [illegible]

[illegible]

Chapter 3
Emergence of Geographical Issues Due to Climate Change

3.1 Geographical Changes and the NWP

Canada has been quite reluctant for some twenty-three years, ever since the transfer of British territories and possessions in North America not already included within the dominion of Canada by Great Britain in 1880.[1] However, from 1903[2] onwards the Canadian government with the intention to consolidate its title to the Arctic islands and its control over the water areas, including those of the NWP inaugurated a number of expeditions at various points along the east coast of the Canadian Arctic Archipelago.[3] In the scholarly writings of Pharand,[4] these expeditions have been summed up as emanating from the intention to take ascendance over the Canadian national security, safety of the Inuits[5] and marine environmental protection.[6] The temptation of other States in the NWP of course lies in the probability of establishing a shorter shipping route connecting the Atlantic and Pacific

[1] Donat Pharand in association with Leonard H. Legault, *Northwest Passage: Arctic Straits*, 1984, Martinus Nijhoff Publishers, p. 38.

[2] *Ibid.*, *Note* that this is the year when Amundsen set out to cross the NWP.

[3] *Ibid.*, p. 39.

[4] Pharand as referred to by the author is the last name of the scholarly writer Donat Pharand.

[5] Donat Pharand, "The Arctic Waters and the Northwest Passage: A Final Revisit", 38 *Ocean Development & International Law*, 2007, p. 3, Accessed via Elin@Lund (date accessed 13 March 2012), where the author explains that the Inuits are Canada's native habitants with their residence in the Arctic region and the effect of continuous shipping in this area could have effects on the Inuits' lifestyle, culture and economy.

[6] *Ibid.*

T. Johansson and P. Donner, *The Shipping Industry, Ocean Governance and Environmental Law in the Paradigm Shift*, SpringerBriefs in Law,
DOI 10.1007/978-3-319-12541-1_3

oceans while sailing between Europe and Asia.[7] In addition, the NWP passage would curtail the original shipping routes through the Panama Canal with 9,000 km and around Cape Horn with 17,000 km.[8] In August 2007, the NWP became admissible to ships without an escorting ice-breaker which was an essential postulate in the *Manhattan* incident.[9] Changes are taking place which is shaping up landscapes as we speak and with that change in mind, coupled with commercial interest, different interest groups are rising to the occasion questioning Canada's claim over the NWP and as to whether the NWP should be acknowledged as an international strait.

3.1.1 The Legal Status of International Straits

A fundamental difference existed as regards the position and definition of "International Straits" between maritime powers and bordering States; while the former insisted an unrestricted freedom of navigation, it was opposed by the latter on the grounds of protecting the marine environment.[10] A general comprehension of this term is laid down by Churchill and Lowe as "…a natural passage or arm of water connecting two larger bodies of water", which is considered as a terse yet compromising definition in the scholarly arena.[11] This push and pull factor as to the determination of an international strait was taken up in the ICJ by Albania and the United Kingdom in the *Corfu Channel* Case where the central issue was whether the North part of the Corfu Channel was to be considered an International Strait.[12] The United Kingdom Government protested to the Albanian Government,

---

[7] *Supra* note 72 in Chap. 2, *Note* the author's reasons which highlight that "[t]here are five to seven different seaways through the Archipelago, including the McClure Strait, the Prince of Wales Strait, and Baffin Bay via the Davis Strait. Except for the route through Baffin Bay and the Davis Strait, the other routes are not suitable for larger ships".

[8] John Falkingham, Dr Humfrey Melling and Katherine J. Wilson, *Shipping in the Canadian Arctic: possible climate change scenarios*, Newsletter of the northern climate change, 2002, p. 4, available at; http://www.taiga.net/nce/resources/newsletters/NCE_Newsletter_Fall2002.pdf (date accessed 13 March 2010).

[9] Hiromitsu Kitagawa, "Arctic Routing: Challenges and Opportunities" in *Impacts of Climate Change on the Maritime Industry*, The proceedings of the Conference on Impacts of Climate Change on the Maritime Industry, 2–4 June 2008 (Sweden), Neil Bellefontaine and Olof Linden (eds.), 2009, World Maritime University Publications, pp. 171–175.

[10] *Supra* note 1, p. 89.

[11] R.R. Churchill and A.V. Lowe, *The Law of the Sea*, 3rd Edition, 1999, Manchester University Press, p. 102.

[12] *Corfu Channel* Case, The Government of the United Kingdom of Great Britain and Northern Ireland versus the Government of the People's Republic of Albania, (Merits), Judgment of the International Court of Justice, 9 April 1949, ICJ Rep. p. 4, where the Albanian Government contended that the sovereignty of Albania was violated because the passage of the British warships on 22 October 1946, was not an innocent passage. The reasons advanced in support of this

stating that innocent passage through Straits was a legal right endorsed by International law and the ICJ firstly concluded that "...generally recognized and in accordance with international custom that states in time of peace have the right to send their warships through Straits used for international navigation between two parts of the high seas without the previous authorization of a coastal state, provided that the passage is innocent".[13] ICJ then proceeded with the question as to whether the strait can be regarded as "International" by highlighting the norms it has to satisfy to enable it to do so and where the decisive criterion was rather its geographical situation as connecting two parts of the high seas and the fact of its being used for International navigation.[14] The "geographical criterion" (connecting the high seas) was transparently determined whereas, in determining the "functional criterion" (International navigation), the Court took into consideration the "number of ships"[15] passing through the Channel during a period of one year nine months. This was a significant strategy adopted by the Court which in turn proved to be important in the legal development of the doctrine of international straits in international conventions. Then again, the North Corfu Channel constituted a frontier between Albania and Greece whereby, a part of it was wholly embraced within the territorial waters of these States.[16]

In the light of these considerations the Court drew the final conclusion that the North Corfu Channel should be considered as belonging to the class of international highways and that it is to be designated as an international strait with an

Footnote 12 (continued)
contention was based on the grounds that the passage was not an ordinary passage, but a political mission; the ships were manoeuvering and sailing in diamond combat formation with soldiers on board; the position of the guns was not consistent with innocent passage; the vessels passed with crews at action stations; the number of the ships and their armament surpassed what was necessary in order to attain their object and showed an intention to intimidate and not merely to pass; the ships had received orders to observe and report upon the coastal defences and this order was carried out. It was however, admitted by the United Kingdom Agent, that the object of sending the warships through the Strait was not only to carry out a passage for purposes of navigation, but also to test Albania's attitude. As mentioned above, the Albanian Government, on 15 May 1946, tried to impose by means of gunfire its view with regard to the passage. As the exchange of diplomatic notes did not lead to any clarification, the Government of the United Kingdom wanted to ascertain by other means whether the Albanian Government would maintain its illegal attitude and again impose its view by firing at passing ships. The legality of this measure taken by the Government of the United Kingdom cannot be disputed, provided that it was carried out in a manner consistent with the requirements of international law. The "mission" was designed to affirm a right which had been unjustly denied. The Government of the United Kingdom was not bound to abstain from exercising its right of passage, which the Albanian Government had illegally denied.

[13] *Ibid.*, p. 28.

[14] *Ibid.*, the two decisive criterion's have been termed as "geographical criterion" and "functional" criterion respectively.

[15] *Ibid.*, p. 29, where the total number of ships estimated was 2,884 where the flags of the ships were Greek, Italian, Romanian, Yugoslav, French, Albanian and British.

[16] *Ibid.*

implicated right of transit passage.[17] Both the "geographical criterion" and "functional criterion" were enshrined simultaneously in Article 16(4) of the GCTS,[18] and the Coastal State jurisdiction was subject to innocent passage.[19] This provision was incorporated in UNCLOS into two separate provisions, i.e. Articles 37 and 44, whereby the definition of international straits is quite verbatim to the definition as laid down in GCTS which states "...straits which are used for international navigation between one part of the high seas ... and another part of the high seas ... ." Although Article 37 embodies both the criteria, a subtle difference is noticed in UNCLOS as regards the "geographical criterion" where a Strait that connects a part of the high seas with a part of the Territorial Sea is not considered to be an international strait. The doctrine of international strait is more extensive in UNCLOS as it provides the rights and duties of Coastal States and Flag States and subscribes to the establishment and maintenance of "navigational and safety aids and other improvements in order to prevent, reduce and control pollution from transiting vessels".[20] UNCLOS covers five categories[21] of international straits and provides for four different legal regimes, which is a compromise as to the types of passage that is permissible in international straits.[22]

---

[17] *Ibid.*

[18] Convention on the Territorial Sea and the Contiguous Zone, 1958, Done at Geneva on 29 April 1958, Entered into force on 10 September 1964, United Nations, *Treaty Series*, vol. 516, p. 205.

[19] Daniel Patrick O' Connell, *The International Law of the Sea*, Vol. I, 1982, Oxford: Clarendon Press, pp. 309–316, where the attempts at the codification of customary international law on international straits made by a number of international bodies such as *Institut de droit* International, the International Law Association and the International Law Commission is highlighted, *See Supra* note 64 in Chap. 2, p. 90, which is a critique to this novel approach and states that these International bodies did not achieve much success as to the "International use" element of the definition and that guidance had to be sought in the *Corfu Channel* Case of 1949 which is still the only International decision on the question.

[20] *Supra* note 5 in Chap. 1, *See* Article 43.

[21] *Supra* note 1, the five categories of international straits include (1) international straits with a route of high seas (or exclusive economic zone) of "similar convenience" and over-flight applicable to the high seas and exclusive economic zone, (2) international straits with a route of high seas (or exclusive economic zone) not of "similar convenience" will be governed by the right of transit passage, (3) international straits with high seas (or exclusive economic zone) route of "similar convenience" seaward of an island (which forms the Strait) of the Coastal State are subject to the right of non-suspendable innocent passage, (4) international straits joining a part of the high seas (or an exclusive economic zone) with the territorial sea of a foreign State governed by the traditional right of non-suspendable innocent passage, (5) international straits joining a part of the high seas (or an exclusive economic zone) which another part of the high seas (or exclusive economic zone) and not included in the previous categories are subject to the right of transit passage.

[22] *Ibid.*, Initially the maritime powers insisted on treating International Straits as an autonomous legal institution, divorced from the territorial sea and providing for the same freedom of navigation and over flight as on the high seas and on the other hand a Coastal State group wished to keep international straits linked to the territorial sea. This would limit the right of passage to one of non-suspendable innocent passage and requiring prior notification or authorization

3.1.2 NWP as Regards to "Geographic Criterion"

In order to satisfy this criterion, the geography of the NWP must link two parts of the high seas and comprise an overlap of Canada's territorial waters. The first element coinciding with the element of "link" is demonstrated in the geographical fact that the eastern end of the passage leads to Baffin Bay, the Davis Strait, the Labrador Sea and the Atlantic Ocean, while the western end leads to the Beaufort Sea, the Chukchi Sea, the Bering Strait and the Pacific Ocean.[23] Nevertheless, comparing with other Straits, the NWP is unique in the sense that it is embraced by ice almost throughout the year and according to Pharand, the Beaufort Sea, which is a part of the Atlantic ocean should not be considered as high seas because of the presence of ice.[24] Considering the veracity of this fact, questions may arise as to whether this presence of thick ice would help designate the NWP even as a regular Strait? Moreover, does this correspond to the "geographic criterion" or does it merely correspond to the restriction and hardship of international navigation which relates to the functional aspect? As is exemplified from the *Manhattan* voyage, there have been vessels, which have transited the passage and icebreakers have escorted all these transits. Article 37 (Part III) read together with Article 38(2) of UNCLOS provides that straits in which there is a right of "continuous and expeditious" transit via exercise of the freedom of navigation between one part of the high seas and another part of the high seas are considered as international straits. These provisions do not provide a bar as to how the vessels should make it through the ice-covered waters and only reflects on the issue of transit. Then again, the pack ice, comprised of ice floes and in constant motion, does not extend to the continental coast of Canada and Alaska for about three months of the year, which does not necessitate the escort of icebreakers.[25] It is assumed that due to climate change the period of ice-existence shall deteriorate expediting the number of vessel-navigation in that part. Even so, if vessels escorted by icebreakers are able to transit the NWP between one part of the high seas to another part of the high seas, then the NWP satisfies the first element of "geographic criterion" of international straits. The second element pertains to an overlap of territorial waters and it

Footnote 22 (continued)
for warships. The compromised article, entitled "Passage of Straits used for International Navigation" was introduced by the United Kingdom at Caracas in 1974 which made two significant changes, the former which substituted the words "the same freedom of navigation as on the high seas" with "transit passage" while the latter limited the application of "transit passage" to straits joining two parts of the high seas and retained the innocent passage rule for straits joining one part of the high seas with the territorial sea of a foreign State. This remained intact in subsequent texts of UNCLOS.

[23] *Ibid.*, p. 99.

[24] Donat Pharand, *The Law of the Sea of the Arctic: with Special Reference to Canada*, 1973, University of Ottawa Press, pp. 174–179, where the suggestion is based on the assumption that the Arctic Ocean, including the Beaufort Sea, is covered by some kind of permanent ice cap which would constitute a complete barrier to navigation.

[25] *Supra* note 64 in Chap. 2.

is estimated that all the NWP routes presupposes passage through such an overlap.[26]

Although Canada has not drawn straight baselines around the Archipelago, it has claimed that the waters of the Archipelago and those of the NWP are internal in nature by virtue of its historic title.[27] Even if there is a supposition that the waters of NWP are not internal, the extension of Canada's territorial waters to 12 nautical miles in 1970 resulted in an overlap of the territorial waters in the western portion on the Barrow Strait.[28] Pharand emphasizes that if the NWP is considered as being only through the M'Clure Strait, then it would still qualify as a legal (territorial) Strait.[29] On the other hand, the route through the Prince of Wales Strait has always had an overlap of territorial waters because the presence of the Princess Royal Islands at about the middle of the strait narrows the strait to less than 6 nautical miles. It seems that prior to 1970 Canada acknowledged that a strip of high seas might exist throughout the M'Clure Strait and the same opinion was held by the U.S. in 1969 when *Manhattan* attempted to cross the NWP by attempting to navigate through it instead of the Prince of Wales Strait. Hence, the NWP constitutes a legal Strait in that it connects two parts of the high seas and has an overlap of territorial waters, which offers the greatest potential for deep-draft navigation.[30]

3.1.3 NWP as Regards to "Functional Criterion"

Some authors are doubtful of the authentication of "functional criterion" of the geographical test of international straits.[31] This contention is also held by interest groups with an intention to dismiss the designation of NWP as an international strait. However, one of the decisive factors that was dominant in the decision of the *Corfu Channel* Case is the "functional criterion", i.e. whether the Strait "…is being used for international navigation".[32] The codification of this criterion in GCTS and UNCLOS provides a strong ground for any strait to undergo this test. Both these conventions spell out the word "used", the concrete interpretation of which indicates the "actual" utility of the strait as opposed to a future anticipated use or a mere prediction of future traffic without any stretch of the imagination.

---

26 *Ibid.*, p. 100.

27 *Supra* note 24.

28 *Supra* note 1.

29 L. Brigham and R. McCalla, *Arctic Marine Shipping, 2009 Report.* Arctic Marine Shipping Assessment (AMSA), *Note* that the NWP consists of five familiar routes or passages amongst which Routes 1 and 2 are considered deep water ones while others have draught restrictions of a maximum 10 m due to the underwater shoals and rocks in the sea.

30 *Ibid.*

31 James Kraska, "The Law of the Sea Convention and the NorthWest Passage" in *International Journal of Marine and Coastal Law*, Vol. 22, No. 2, 9 May 2007, p. 257.

32 *Supra* note 12.

This is parallel to the conclusion drawn by Pharand that "...before a strait may be considered international, proof must be adduced that it has a history as a useful route for international maritime traffic".[33] The "functional criterion" is dependent on two elements as has been observed in the *Corfu Channel* Case where the former involves the number of vessels that have had transit during a certain period and the latter responds to the number and types of Flag States, which are represented during those transits. Pharand is of the opinion that the NWP does not satisfy the "functional criterion" and concludes, "[t]hroughout the 80-year history of attempted exploratory navigation in the NWP, only 40 complete transits of the passage have taken place and of these 27 were Canadian ships" and forwards a statistical explanation that "[a]mong the 13 foreign crossings, 10 were American, 1 Norwegian, 1 Dutch and 1 Japanese".[34] However, the Canadian Coastguard has estimated that after the completion of the first transit in 1906, there has been 69 non-Canadian transits in the NWP and during the year 2004–2005, there were only 7 transits.[35] This when compared to the number of transits and Flag States of the *Corfu Channel* Case is relatively low.[36] Other scholars like McDorman[37] and Mckinnon[38] have had similar views of Pharand and argue that the number of transit in the NWP is insignificant and that it does not meet the "functional criterion" due to the fact of the low number of international transits. However, Rothwell is of a different opinion and states that "[c]ertainly, the amount of traffic through the NWP is not comparable to that of the *Corfu Channel*, or other commonly accepted international straits. The need to apply different standards in the polar regions, however, has been recognized".[39] He also argues that the presence of icc in the Passage and the polar weather conditions should allow for a test requiring a lower volume of international navigation of the passage in order to classify it as an international strait.[40] However, according to Pharand, a review of shipping activities within the NWP clearly exemplifies that the "functional criterion" even assessed according to a polar standard, has not been satisfied. Nevertheless, it seems that the two international conventions which codify the international strait regime do not exactly provide a general threshold as to the number of vessels and the number and

---

33 *Supra* note 5.

34 *Supra* note 1.

35 Canadian Coast Guard, *The Northwest Passage: Summary of Complete Transits, Fleet News no. 11*, 1985.

36 *See Supra* note 1.

37 Ted L. McDorman, "In the Wake of the 'Polar Sea': Canadian Jurisdiction and the Northwest Passage", *Les Cahiers de Droit*, 1986, Vol. 27, pp. 623–646, where the author states that "The Passage is not a crucial international thoroughfare, it has limited strategic importance, it is used almost exclusively by Canadians ...".

38 J.B. McKinnon, "Arctic Baselines: *A Litore Usque ad Litus*", in 66 *Can. Bar Rev*, 1987, pp. 790–797.

39 Donald R. Rothwell, "The Canadian-U.S. Northwest Passage Dispute: A Reassessment", 26 *Cornell Int'l L.J.*, 1993, pp. 331–352.

40 *Ibid.*

types of Flag States that need to ply back and forth for transit in the NWP. There is a vacuum of case-law in this regard and it is only a matter of time before this issue is placed before an International Tribunal to indicate a specific figure which will help determine the "functional criterion". It can under no circumstances be adjudged on the basis of comparison with a higher figure since the environmental aspects of it must be duly considered. International navigation will surely internationalize the NWP and if the "potential use" and the lack of a specific threshold are taken into account, then the NWP has clearly satisfied the "functional criterion".[41] The scholars must take into consideration the issue of global warming and the alarming rate at which the Arctic ice is melting, which is a catalyst to the internationalization of NWP and with that, the number of transits are likely to increase.

3.1.4 Sector Theory and the Canadian Arctic

Linked principally with the Antarctic, the Sector theory, which is a method for delimiting an area claimed by a state, has had a pivotal role in international law as a basis of claiming jurisdiction in the Arctic waters by Canada.[42] This claim of jurisdiction has become more concrete as the Arctic surroundings are shaping up due to global warming, rendering Canada a reason to be more resistant in its claims on adjacent waters. The Russian Federation has also applied the Sector theory as a basis of their land and maritime claims.[43] Initially what started out as a

---

[41] *See Supra* note 37, where McDorman speculates that promises of mineral wealth in the Arctic will doubtless lead to increased maritime traffic, *See* also *Ibid*, where Rothwell points to evidence of an already increasing usage of the Passage of twenty-three transits, eight by non-Canadian flagged vessels recorded during the 1980s alone, *See Supra* note 24, where Pharand shares this view, warning that International navigation has already begun in the eastern part of the Passage, used for the transportation of minerals from the Nanisivik Mine to the south of Lancaster Sound, and the Polaris mine, north of Barrow Strait.

[42] Suzanne Lalonde and Ronald St. J. MacDonald, "Donat Pharand: The Arctic Scholar" in *The Canadian Yearbook of International Law-Annuaire Canadien de droit international*, 2006, Vol. XLIV, University of British Columbia Press, p. 59, where the authors have cited from Pharand which states "[t]he Sector theory has been invoked by a number of politicians and officials in Canada as a legal basis for claiming jurisdiction not only over the islands of Canadian Arctic Archipelago, but also over the waters within and north of the islands right up to the pole".

[43] *Infra* note 108 in Chap. 2, where the author explains that internationally, the Soviet Union has adopted the Sector theory, at least in literature, where Soviet writers refer to the Soviet northern seas as the entire Soviet sector up to the North Pole. This sector of the Arctic is considered as historical waters to which the Soviets have a historical right, created by the Russian people's industrious work during several centuries. A second Russian reference enclosing all of the Arctic up to the North Pole is contained in a note to the American Government following a 1924 incident. A decree in 1926 stated that the treaty between the U.S. and the Russian Federation, which draws a demarcation line between the two countries in the Bering Strait, "goes in a direct line to the north, without deviation right to the Polar Sea". Using this line as the eastern sector line the Soviets also propose a western line that would go from 32°04′35″ East to the North Pole with a deviation to allow for the Norwegian-Russian agreement on the Svalbard Islands.

claim over the Arctic Archipelago is now being applied to the EEZ and Continental Shelf. According to Pharand, the Sector theory in the Canadian Arctic is attributed to Senator Pascal Poirier, in 1907, who was a pioneer "to actually systemize the use of meridians of longitude to claim territorial sovereignty in the Arctic".[44] Senator Pascal Poirier asserted that Canada owned everything within a pie-shaped sector extending from the continental coastline to the geographic North Pole.[45] Three elements are integral parts of this theory and "Boundary treaty", as the first element of three legal foundations has been dismissed by Pharand on the ground that the *1825*[46] and *1867*[47] boundary Treaties cannot serve as a legal basis as in both instances the subject matter was land only, and not land and sea, except for inland waters and the territorial sea.[48] Pharand's evaluation of the second element, the doctrine of "Contiguity" as a legal foundation is held to be threefold, i.e. *state practice,*[49] *International decisions*[50] and *doctrinal opinion*[51] and has been disregarded as insufficient insofar as they cannot serve as a legal basis for the acquisition of territorial sovereignty. The sole reason was that, since "Contiguity", by itself, is incapable of generating sovereignty over land areas, it cannot be justified as an instrument of acquisition over the sea areas.[52] The final legal foundation of the Sector theory, according to Pharand, lies in the practice of "Customary law". Then again, according to Pharand, Canada has never adopted any legislation in Council claiming a sector and has resorted to a number of official steps indicating its reliance on it which is inconsistent in pattern and position.[53] Boundaries following sector lines have evolved in maps since 1952 and this method has been

---

[44] *Ibid.*

[45] Robert S. Reid, "The Canadian Claim to Sovereignty Over the Waters of the Arctic", 12 *The Canadian Yearbook of International Law*, 1974, pp. 111–115.

[46] *Treaty of Saint Petersburg, 1825.*

[47] Convention between the U.S. and His Majesty the Emperor of Russia, for the Cession of the Russian Possessions in North America to the United States, Concluded at Washington, March 30, 1867; Ratification Advised by Senate, April 9, 1867; Ratified by President, May 28, 1867; Ratifications Exchanged at Washington, June 20, 1867; Proclaimed, June 20, 1867.

[48] Donat Pharand, *Canada's Arctic waters in international law*, 1988, Cambridge University Press, pp. 17–25.

[49] *Ibid.*, p. 38 where Pharand considers the United State of Americas' reliance on the doctrine in the nineteenth century, that of various European powers with respect to colonial Africa, as well as that of Canada and the Russian Federation with respect to the Arctic.

[50] *Ibid*, p. 38, where Pharand discusses the *Aves Island Case* (1865), the *Island of Bulama Case* (1870), the *British Guiana Case* (1904), the *Island of Palmas Case* (1928), the *Eastern Greenland Case* (1933), the *Minquiers and Ecrehos Case* (1953) and the *Western Sahara Case* (1975).

[51] *Ibid.*, p. 40, where the doctrinal opinions as discussed by Pharand include the French jurist Fauchille, the American jurist D.H. Miller, the English writer M.F. Lindley, the Soviet writer W.L. Likhtine, the British Professors C.H.M. Waldock and H. Lauterpacht.

[52] *Supra* note 42.

[53] *Ibid.*, p. 77.

accounted for as part of the subsequent conduct of the Parties as an aid for the interpretation of a treaty.[54] But the involvement of treaty is impertinent in this regard to determine the eastern line of the sector. The western line corresponding to the meridians of longitude were used only to delimit the territorial possessions of the parties and the 1825 and 1867 boundary treaties were already adjudged by Pharand, to have no legal basis.[55] Hence, regardless of the probative value of maps and charts in certain circumstances, it would not provide Canada with any logical support in favour of its own claim. The U.S. has never relied on the pragmatic aspect of the Sector theory and has expressed opposition in its usage as a basis for claiming sovereignty over the Arctic Ocean. It is of the view that division under the Sector theory would in fact constitute claims of sovereignty over the high seas and this novel approach to artificially create a closed sea would in repercussion infringe the rights of all nations to the free use of the area.[56]

3.1.5 A Legal Alibi Distorting International Regime?

As Pharand suggests, Canada's claim to the waters as internal would be supportable if reliance was placed on the 1951 decision of the ICJ in the *Anglo-Norwegian Fisheries Case*[57] by which Canada's Arctic could be equated with the islands along the Norwegian Coast via use of straight baselines.[58] Under the domain of customary international law, as applied by the ICJ in this case, there exists no right of passage in waters enclosed by straight baselines, regardless of the previous status of the newly enclosed waters.[59] This was subsequently modified in the GCTS which made the enclosed waters subject to the right of innocent passage on the condition that they were previously territorial waters or high seas. Since Canada is not a party to GCTS, it had to rely on customary law for the validity of its straight baselines which as a result is not secure since the Convention provision itself was a part of customary law.[60] A further investigation would illuminate the fact that the GCTS provision for innocent passage in newly enclosed internal waters has had universal usage as to become legally binding on all States. Then again, the claim

---

[54] *Ibid.*

[55] *Ibid.*

[56] Bo Johnson Theutenberg, *The Evolution of the Law of the* Sea, 1984, Tycooly International Publishing Company.

[57] *Anglo-Norwegian Fisheries Case* (United Kingdom vs. Norway), 18 December 1951, I.C.J. Report, p. 116, *See* also L. C. Green, *The Modern Law Review*, Vol. 15, No. 3 (July, 1952), pp. 373–377.

[58] Donat Pharand, "The legal regime of the Arctic: Some outstanding issues", *International Journal,* 1984, Vol. 34, No. 9, p. 748–769.

[59] *Supra* note 48, p. 228.

[60] *Ibid.*

of Arctic as historic internal waters by reason of straight baselines could also be expedited if reliance could be placed upon Inuit use of the waters and surrounding land owing to the fact that it would indicate a presence over an extended period of time.[61] But this technicality was avoided by Canada because of the significant issues that would arise for the Government in its negotiations with the Inuit over outstanding land claims.[62] Moreover, Canada has never publicly asserted its use of straight baselines during the *Polar Sea* voyage in 1985 although there existed an agreement of Continental Shelf boundary founded on straight baselines between itself and Greenland in 1973.[63] A diplomatic protest would have certainly justified Canada's position in the NWP. Nevertheless, peer pressure from the international community, more specifically the U.S. for Straits to embody the right of International navigation has been consistent even prior to the *Manhattan* voyage.[64] More convincing is the recent trend of international shipping across the NWP which has developed over relatively few years. Pharand infers that "[o]n the question of international shipping, it has already begun in the eastern part of the passage…" and further adds that "[a]lso, it seems to be only a question of time before regular shipping takes place…along the full length of the Northwest passage".[65] Assuming this to be correct, now the obvious question is what right of passage would apply after internationalization? As is understood, a consensus has been reached in UNCLOS as to the type of passage that is applicable in the Straits termed as "transit passage". This term was advocated by the U.S. and the Russian Federation to act as a substitute for the term "free passage".[66] This definition contains two limitations where the first one is to exercise the given "freedom" in accordance with this part and the second one is that transit must be "continuous". This "continuous" transit is in conjunction with the term "expeditious" where the burden of proof lies with the defendant. These certain limitations have been embedded in the form of duties imposed on ships and aircraft during transit in Articles 39, 40 and 41.[67] This is as it exists today, is what is incorporated in theory. In practice, however, the NWP is for the moment not an international strait and consequently the right of transit passage is absent. Consent of Canadian authorities is necessary for American ice-breakers to enter NWP, which is

61 *Supra* note 37, p. 634.

62 *Ibid.*

63 *Ibid.*

64 J. M. Spinnato, "Historic and Vital Bays: An Analysis of Libya's Claim to the Gulf of Sidra", 13 *Ocean Development and International Law Journal*, 1983, pp. 65–68, where it is stated that, as regards navigational rights the United States has persistently taken a strong stand to safeguard the right of navigational passage, and in August 1981, the U.S. aircraft and Libyan aircraft were involved in a brief skirmish in the Gulf of Sidra that Libya had declared in 1973 to be an historic bay and hence internal waters in which the U.S. vessels were not permitted.

65 *Supra* note 1, p. 230, para 2.

66 *See* 1971 Draft Articles of the United States, A/AC.138/S.C.II/L.4 and the 1972 Draft Articles of the Soviet Union, A/AC.138.S.C.II/L.7.

67 *Supra* note 5 in Chap. 2.

incorporated in the agreement of 1988 and is still applicable.[68] Whether there exists a right of innocent passage in the NWP, according to customary law, seems to be controversial. McDorman is of the opinion that if an international tribunal was looking at the legal cases of Canada and the United States at the time the *Polar Sea* traversed the NWP, the tribunal would reject the Canadian position and would similarly reject the position of the U.S. He further adds that the tribunal would be left with the option that the NWP was part of Canada's territorial sea in which foreign vessels would have the undisputed right of innocent passage. Then again, the applicability of Sector theory to Arctic waters has received International critique. The Canadian government's support of the Sector theory as a means of defining the EEZ and continental shelf boundaries can only be seen as increasing. This view of the Sector theory does not conflict with the earlier view held by the U.S. that the Sector theory could not be used to enclose the high seas as only the zones of UNCLOS would be defined with this principle. International navigation for the shipping industry is significant to uphold the commercial pillar which is a part of international commercial law. But where this right of transit passage is subject to objection by controversial theories, it can be held that those theories constitute a legal alibi which is distorting the international regime. Different interest groups have laid down different claims as a result of climate change which is shaping up the adjacent sea areas. Sovereign claims must be coupled with the flexibility as endorsed by international conventions to give due regard to innocent passage of Flag States. If otherwise, it leads to the contortion of the international regime. The ultimate question is whether the region will develop in an orderly fashion based on the existing set of International rules that have been primarily established by UNCLOS or, will the circumpolar States increasingly turn to unilateral actions based on random theories which will face challenges from other States and international bodies for control. The answer is not readily available until and unless the fact of matter has been placed before the international courts.

3.2 Compromise in the NSR?

In theory, it is claimed that distance savings through NSR can be elevated compared to the currently used shipping lanes via Suez or Panama.[69] The history trails back to the time when the English colonial powers expanded their empires and trading routes into Asia in the sixteenth century in search of a shorter sea route.[70]

---

68 *Agreement between the government of Canada and the government of the U.S. on Arctic cooperation* (Canada-U.S.), 11 January 1988, C. T. S. 1988/29, available at; http://www.lexum.umontreal.ca/ca_us/en/cts.1988.29.en.html (date accessed 20 March 2012).

69 *Supra* note 42 in Chap. 2.

70 *Ibid.*, In theory the set of navigational routes between Kara Gate in the west and Bering Strait is known as the Northern Sea Route which corresponds to a number of narrow straits which ultimately represent a constraint in safe sailing.

The legal regime of the waters North of the Soviet coastline between Norway and the Bering Strait has been consolidated and transformed significantly since the Soviet government took steps to become signatory to UNCLOS.[71] Then again, this transformation was coupled with collateral developments potentially concerning the utilization of Arctic sea lanes and the exploitation of Arctic resources which included the continuous use of the Arctic waters by several States to reach the North pole, joint resource exploitation and joint shipping on the Soviet Arctic continental shelf and most importantly the utilization of the NSR as an international shipping route between the Atlantic Pacific Oceans.[72] Among the four maritime boundary agreements, the first maritime boundary in the Arctic was negotiated between Norway and the then Soviet Union in 1957.[73] The Russian Federation favored a boundary that coincided with a Sector line as applied in its national legislation and rejected the application of equidistance based on special circumstances.[74] On the other hand, the longest maritime boundary was successfully negotiated between the U.S. and the then Soviet Union in 1990 which established the boundary for the territorial sea, EEZ and the continental shelf in the Arctic Ocean, Bering Sea and the Chukchi Sea respectively and relevantly, the Russian Federation continental shelf submission[75] complies with this treaty so as not to define any portion of the Arctic continental shelf to the East of the boundary.[76] The rejection of this submission has more or less instigated the Russian Federation to develop an introvert Arctic policy owing to the prediction that climate change would surely internationalize the NSR. However, compliments may be rendered to these MOUs as it reflects the commitments of the States concerned with the NSR to retain a good relation with their Arctic neighbours. Nevertheless, this appearance has taken its toll when the cooperation has been disrupted following the

[71] William E. Butler, "The legal regime of Soviet Arctic marine areas" in *The Soviet Maritime Arctic*, Lawson W. Brigham (ed.), 1991, Belhaven Press, p. 215, para 1, *Note* that the author explains the alteration which "…takes the form chiefly of legislation giving effect to the 1982 LOS".

[72] *Ibid.*, p. 216.

[73] Brian J. Van Pay, "National Maritime Claims in the Arctic" in *Changes in the Arctic Environment and the Law of the Sea*, Myron H. Nordquist, John Norton Moore and Tomas H. Heidar (eds.), 2010, IDC Publishers, Martinus Nijhoff Publishers and VSP, p. 71, *Note* that the author refers to the *Agreement between the Royal Norwegian Government and the Government of the Union of Soviet Socialist Republics* concerning the sea frontier between Norway and the U.S.S.R in the Varanger Fjord, signed 15 February 1957.

[74] *Ibid.*, *Note* that the competing claims resulted in a series of disputed areas totaling 175,000 km. *See* also Alex G. Oude Elferink, "Arctic Maritime Delimitations: The preponderance of Similarities with Other Regions" in *The Law of the Sea and Polar Maritime Delimitation and Jurisdiction*, Alex G. Oude Elferink and Donald R. Rothwell (eds.), 2000, 185 Kluwer International, Great Britain.

[75] The Russian Federation made a submission to the Commission on the Limits of the Continental Shelf, pursuant to Article 76 of UNCLOS on 20 December 2001.

[76] *Ibid.*, p. 74.

Russian Federation's placement of a flag at the North Pole.[77] Although this issue dissolved with time, it left a question on the face of mutual agreement which may be threatened when opposing groups place unilateral "land-grabbing" claims as the NSR witnesses the dawn of climate change.

Future forecasts have been laid down as to the emergence of seasonal lanes in the NSR through the ice locked Arctic.[78] This trend of climatic change if constant would result in the disappearance of summertime ice caps and this significant retreat of ice cap will ensure the possibility to use the NSR in shipping which would eventually reflect less trip distances and hence, a reduction in the operation cost. As the framework conditions for the NSR is undergoing a sloth alteration, the Russian Federation petroleum activities are moving northwards and is expected to go offshore.[79] This aggrandized petroleum activity will lead to unprecedented levels of shipping in and westwards from the Barents and Kara seas.[80] If the Russian Federation can offer competitive considerations, there might be individual ship-owners willing to make sporadic use of the transit route which depends on the foreign policies it adopts. Moreover, on the issue of continental shelf, the government of the Russian Federation adopted a new Arctic strategy in September 2008 which emphasizes the region's importance to the Federation's economy as a major source of revenue.[81] The national security strategy released by the Russian Federation indicated a possible conflict in the future over the question of rights over Arctic oil coupled with the possibility of war breaking out as a result of the struggle for control thereby.[82] This strategy is inferred to be congruous to most foreign objectives that relate to the opening of the NSR for very limited period of the year which could be accessible for longer periods in the future. However, compliments may be given to the Russian Federation's acquiescence in observing UNCLOS before adoption of any new International agreement relating to the coordination in using the Arctic spaces.[83] It may be deduced that the reason behind this conformity lies in the Russian Federation's intention to prevent distortion of the international regime and to tone down the line of hostility with its Arctic shareholders. Although the Arctic strategies reveal its efforts to use special forces to establish its jurisdictional claims, on the other hand, it has indicated its positive face to cooperate via existing MOUs for further exploitation of resources in the NSR.

---

[77] R. Hubert, "Cooperation or Conflict in the Arctic" in *Changes in the Arctic Environment and the Law of the Sea*, Myron H. Nordquist, John Norton Moore and Tomas H. Heidar (eds.), 2010, IDC Publishers, Martinus Nijhoff Publishers and VSP, p. 32.

[78] *Supra* note 71.

[79] *Supra* note 42 in Chap. 2.

[80] *Ibid.*

[81] Alexander S. Skaridov, "Russian Policy on the Arctic Continental Shelf" in *Changes in the Arctic Environment and the Law of the Sea*, Myron H. Nordquist, John Norton Moore and Tomas H. Heidar (eds.), 2010, IDC Publishers, Martinus Nijhoff Publishers and VSP, p. 490.

[82] *Ibid.*

[83] *Ibid.*

Chapter 4
Emanation of Environmental Issues and Regional Responses

The level of International interest in the Arctic has surely intensified. It has already been established that, with this interest flows the increased action of international navigation that seeks commercial concern through a shorter sea-route. The first norm of the result of climate change is the geographical issue while the final norm relates to the environmental issue. To be precise, climate change is instigating changes in the legal regime. Once the "land-grabbing" claims are resolved and the right of innocent passage has been granted, the Arctic States will have to give due consideration to the problems that arise out of International navigation i.e. the environmental issues. A single major oil tanker casualty in the Arctic may have serious environmental outcomes. The environmental issue in this context can vary from global, regional, trans-boundary, domestic or an amalgamation of these.[1] In order to examine environmental law in an international context, it is important to observe the interplay of International, regional, sub-regional and national rules and institutions.[2] Questions have been raised as to whether the region will respond to the environmental changes in an orderly fashion based on the existing set of international rules that have been primarily established by UNCLOS i.e. Flag State, Port State and Coastal State jurisdiction, or will the circumpolar States increasingly turn to unilateral, bi-lateral or multilateral actions in an effort to safeguard the marine environment.[3] The answers for fact lie in the *status quo* efforts initiated at the circumpolar level. A critique of these efforts would certainly highlight the fact as to whether they have instigated a modification in the legal regime and to what extent these endeavours have been able to respond to the environmental changes.

---

[1] *Supra* note 46 in Chap. 1, pp. 6–7.

[2] Oran R. Young, *The Structure of Arctic Cooperation: Solving Problems/Seizing Opportunities*. Fourth conference of Parliamentarians of the Arctic Region, Rovaniemi, 2000, available at; http://www.arcticparl.org/_res/site/File/images/conf4_sac.pdf (date accessed 30 March 2012).

[3] *Supra* note 77 in Chap. 3, p. 30, para 2.

T. Johansson and P. Donner, *The Shipping Industry, Ocean Governance and Environmental Law in the Paradigm Shift*, SpringerBriefs in Law,
DOI 10.1007/978-3-319-12541-1_4

4.1 Arctic Council Initiatives

It was not until the end of the Cold War that the initial effort to form multilateral agreements began to succeed.[4] As the Soviet Union began to transform, the Finnish Government sought for means of promoting cooperation which led to the creation of the AEPS which ultimately transformed into the Arctic Council.[5,6] Although the AEPS has been assimilated in the infrastructure of the Arctic Council, till date it remains as a valid strategy of working groups for the Council. Keeping a constant focus on the international environmental issues, the Council has moved to strengthen relations among the Arctic States. Then again, it has facilitated the means of obtaining a mutual understanding of the environmental challenges that the Arctic States are facing.[7] The Arctic Council is the directing authority of the Arctic legal regime. However, it is not considered as an international organization with a legal personality and describes itself as a "high-level forum intended to provide a means for promoting cooperation among Arctic states… on common Arctic issues, in particular issues of sustainable development and environmental protection in the Arctic."[8] But the Council underwent the same frustration which plagued the AEPS as limited resources hindered the development of an International yet effective strategy. All of the working groups were retained, and the task force on sustainable development was transformed into a working group. Hence, the Arctic Council may be distinguished from AEPS insofar as the Council is specifically mandated to develop a sustainable development program.[9] The Arctic is transforming into a geopolitically important factor and the growing interest in Arctic issues stem from

---

[4] *Ibid.*

[5] *Note* that the Arctic Council is created by the Ottawa Declaration 1996.

[6] David Scrivener, *Environmental Cooperation in the Arctic: From Strategy to Council,* Security Policy Library No. 1, (Oslo: Norwegian Atlantic Committee, 1996), *Note* also that the AEPS divided its activities between working groups and task forces. The four working groups were the Arctic Monitoring and Assessment Program (AMAP); Protection of the Arctic Marine Environment (PAME); Emergency Prevention, Preparedness and Response (EPPR), Conservation of Arctic Flora and Fauna (CAFF). However, because the AEPS was not supplemented with its own sources of funding and was not a formal treaty, its capacity to provide remedies for the environmental problems it uncovered was to a certain extent restricted. It was these restrictions that paved the way to the creation of the next major environmental arctic initiative i.e. the Arctic Council.

[7] *Supra* note 77 in Chap. 3, where it is stated that the Council compels the senior leaders to focus on Arctic issues every two years when ministerial meetings are held.

[8] Linda Nowlan, *Arctic Legal Regime for Environmental Protection*, 2001, IUCN, Gland, Switzerland and Cambridge, UK and ICEL, Bonn, Germany, available at; http://weavingaweb.org/pdfdocuments/EPLP44EN.pdf (date accessed 31 March 2012).

[9] David VanderZwaag, R. Huebert and Stacey Ferrara, *The Arctic Environmental Protection Strategy, Arctic Council and Multilateral Environmental Initiatives: Tinkering while the Arctic Marine Environment Totters*, Denver Journal of International Law and Policy, Vol. 30:2, pp. 131–171.

climate change, melting ice-caps and improved access to the Northern regions rich in oil, gas and minerals. It is argued that the interest in the Arctic issues is a privilege for the Arctic nations since many of the problems found in the Northern environment are the result of activities of non-Arctic States. Then again, it is believed that the Arctic Council can provide the basis for an international platform on environmental policy. But a significant drawback is that the main course of action for the Arctic Council has been identifying environmental threats and establishing Guidelines for a course of remedial action, while leaving the duty of implementation to the individual Arctic nationalities. The Arctic States are responsible for responding according to its own means. In practice, the disparate working groups that have been specifically mandated to deal with marine environmental issues, have not so far developed a mandate that goes beyond determining the marine environmental issues facing the Arctic region.

4.2 Analysing OSPAR: Arctic Regional Instrument on Dumping of Wastes

OSPAR is a regional instrument which aims to amalgamate the Oslo Convention and the Convention for the Prevention of Marine Pollution from Land-based Sources (Paris Convention), whereby Annex II deals with prevention and elimination of pollution by dumping or incineration.[10] Although OSPAR is a regional instrument, the preamble acknowledges national, regional and International actions and calls for a combined effort when dealing with marine pollution. Then again, this Convention expressly includes a definition of internal waters and embodies such waters within maritime area where the convention applies.[11] Although OSPAR addresses two primary concerns corresponding to Ocean dumping of radioactive wastes and the dumping of dredged material, the significant drawback of the convention is that it covers only a part of the Arctic marine environment. Since OSPAR has membership different from the Arctic region, countries with no national interest in the Arctic region may influence in decision making not unique to specific Arctic concerns.[12] The three Arctic non-signatories possess jurisdiction over an extensive regime of the marine environment and a complex situation may arise where these States would separate the Arctic regime in their domestic environmental legislation. Moreover, it relies on the domestic administrations to prevent and punish conduct that contravenes the convention. Since Denmark and Norway are the

---

[10] Official Webpage of OSPAR, available at; http://www.ospar.org/content/content.asp?menu=00340108070000_000000_000000 (date accessed 27 February 2012).

[11] *Ibid.*

[12] *Ibid.*, p. 105.

only Arctic States which are parties to this convention, it is impossible to implement the provisions or policies of this regional instrument on non-contracting States. In addition, there is no means of monitoring the domestic administrations for strict compliance of its regulations and as a result protective measures cannot be ensured against illegal dumping in the Arctic waters.

4.3 Regional Response as a Catalyst of Change

Although the Arctic has not witnessed the development of competent and persuasive multilateral agreements, the Arctic Council has managed to address certain significant issues. Nevertheless, it remains limited as an instrument which is dependant on multilateral agreement. The provisions of international law is the first layer of the Arctic legal regime and UNCLOS does not deal with marine environmental issues elaborately but only gives reference to GAIRAS. A traditional interpretation of UNCLOS wording and the definition of pollution would indicate the fact that climate change was not the original intention of the policy makers.[13] Moreover, there is no indication that the parties were intending to limit the definition to those threats exclusively, that were identified at the time of negotiating UNCLOS.[14] It is relevant to mention that the Polar Shipping Guidelines which also acts as an International instrument only provides for environmental protection incorporated in a single part and is characterised as insufficient, brief and non-legally binding in nature. On the other hand, it is not always the fact that national legislation corresponds to the International systems verbatim and the Arctic being a region of marine diversity, it would be impossible for the Arctic States to enact similar national legislation. These dissimilar legislations may instigate conflict of laws and this is where the regional responses come into play. Hence, the emergence of regional cooperation although it relies on the generosity of its member States to provide the resources needed to engage in its activities.[15] The "soft law" approach of regional cooperation takes its position as a legal regime in the Arctic domain. This regional response is in fact indicated as a modification to the legal regime in its entirety since the general focus is on international law and the national legislation that follow. The Arctic States have committed to engaging in a circumpolar agreement where the Arctic Council is the flagship of this dimension of legal regime. Although it is not entrenched as "hard law" through a treaty and defined on the basis of its flexible influence, it is in reality envisioned as a catalyst of change in the legal domain of the Arctic.

---

[13] Meinhard Doelle, *Climate Change and the Use of the Dispute Settlement Regime of the Law of the Sea Convention*, Ocean Development International Law, 2006, Vol. 37, Issue: 3–4, Taylor & Francis Publishers, pp. 319–337.

[14] *Ibid.*

[15] *Supra* note 77 in Chap. 3, p. 40.

4.4 A Critique of the "Council"

Many of the Arctic Maritime boundaries remain disputed and the five Arctic States have been unable to resolve any of the existing maritime disputes.[16] Accepting the veracity of this statement, it can be deduced that the Arctic Councils endeavours in trying to establish MPAs within the zones of a specific Coastal State or to adopt special Resolutions for a particular area cannot be accomplished due to those overlapping claims. After pinpointing coordinates of those boundaries the individual States can relate to the guidelines or resolutions suggested by the Council and define "special areas" to protect and preserve their designated part of the Arctic with stringent legislation. Till then, it is just a "wait and see" policy. Then again, contemporary international environmental law is aimed at regulating emerging environmental issues by setting common international standards for prevention or mitigation of harm. It also aspires to provide an adaptable rule-making technique that can allow for flexible and regular amendments and much of this regulatory system is composed not only of multilateral treaties but also "soft law" techniques provided by regional organizations. In this regard, it is submitted that the Arctic Council has to a greater extent left a trail of success. Nevertheless, with the entrance opening up in the Arctic, the Council needs to be alert as regards any contamination of the marine system and the fact that the Arctic requires a "multiple user" MPA in Coastal and offshore environments resulting in that a single MPA can comprise a mosaic of management coupled with restriction categories in terms of voluntary marine-pollution. Taking this into account and to designate 'any marine area' in the Coastal or offshore environments and areas beyond national jurisdiction as MPA there needs to be a development of a circumpolar network in the Arctic which is complicated because of the mix of national and international agencies and jurisdictions with responsibilities in the region. This responsibility lies in the wake of the Arctic Council and is yet to be on its agenda. In addition, the very structure of the Council's working arrangement, divided as it is among working groups, places limits on the ability of the Council to clasp with the complex and interrelated problems posed by the new development opportunities in the region.[17] Then again, the Arctic Council could reflect a determinant role by acting as the high-level forum it intends to be and becoming the coordinating nexus for the Arctic legal regime. However, for this approach to work, there remains the need of a significantly strengthened Arctic Council where the question of funding must be prioritised. Otherwise the Council could never rise to the occasion to act as a decisive force. The Council should be the balancing factor in the region which can amend guidelines to tackle any new form of environmental hazards caused by the increased

---

[16] *Ibid.*, pp. 41–43.

[17] R. Hubert and Brooks B. Yeager, *A New Sea: The Need for a Regional Agreement on Management and Conservation of the Arctic Marine Environment*, January 2008, A Report Published by WWF International Arctic Programme, Oslo, Norway, p. 23.

international navigation. The ramified working groups of the AEPS and the Arctic Council could also be upheld and given extended mandates and more resources in dealing with multifarious environmental issues at large which in the near future may emerge with the dawn of climate change. The analysis as mentioned is in fact a theory. The practicality of it lies with the individual States who are participants of this Council and who need to assure primarily, that the existing boundary agreements remain constant. The Council needs to be given that "designated area" to implement its policies. The Arctic Council needs to shift gear to being an organization that has much more of a presence in domestic and international affairs and such a shift should give buoyancy to issues like solving maritime boundary problems and then dealing with complex issues of the marine environment. Then and only then will the enactments of the Arctic Council manifest a high level of sophistication and competency. To what extent can the "soft law" of the Council succeed, remains to be seen.

Chapter 5
Strict Ocean Governance and Commercial Implications of Arctic Navigation

There has been large-scale ocean-based international trade for millennia.[1] The Maritime business is considered an antiquated yet most economic and intensive business coupled with a high economic sensitivity due to its global nature. Ocean-based international trade continues to expand and there are several current trends that are driving the continuing increase in scale.[2] Gradual liberalization of international trade and liberalization of international finance since World War II have acted as accelerators of this continuous expansion.[3] Climate change, in addition, entices States to engage in geographical distribution of the oceans and develop international trade to accommodate large-ship traffic which has commercial implication. This gives rise to ocean governance whether international or national which, in turn, has an adverse effect on global trade. The Arctic will not be an exception to this trend. The relationship between Arctic international trade and Ocean governance is complicated and needs to be justified.

[1] Elizabeth R. DeSombre and J. Samuel Barkin, "International Trade and Ocean Governance" in *Securing the Oceans: Essays on Ocean Governance-Global and Regional Perspectives*, Chua Thia-Eng, Gunnar Kullenberg and Danilo Bonga (eds.), January 2008, Published by GEF, UNDP and IMO in association with the Nippon Foundation, p. 160, where it is stated that such trades include the trade between ancient Rome and Egypt, and between Mediterranean and the Baltic from the late Middle ages to the Renaissance.

[2] *Ibid*, *See* also, Martin Stopford, *Maritime Economics*, 2009, London: Routledge, Taylor and Francis Group, where it is highlighted that "[t]he transport industry has been one of the prime forces responsible for shifting the world from an essentially national system to the global economy that exists today."

[3] *Ibid*. pp. 161–163, where it is stated that "[l]iberalization over the past 2 decades has had the effect of bringing countries into the international trade system, such as China, that were previously only marginally involved in it".

T. Johansson and P. Donner, *The Shipping Industry, Ocean Governance and Environmental Law in the Paradigm Shift*, SpringerBriefs in Law,
DOI 10.1007/978-3-319-12541-1_5

5.1 Troubled Waters of Arctic and Economic Incentives

It is already substantiated that the NSR has an indisputable advantage in travel between Europe and the Far East or the West Coast of North America. The navigational distance between Hamburg and Yokohama, through NSR is 6,920 nautical miles when compared with travelling via Suez Canal which covers 11,439 nautical miles while going around the Cape of Good Hope.[4] At the same time the NWP is a shortcut between the East Coast of North America and its West Coast or the Far East which is about 9,800 nautical miles in comparison with 12,420 nautical miles via Panama Canal.[5] Curtailing longer distances has been a constant practice for stakeholders of the shipping-industry since cargoes are delivered expeditiously by consumption of less fuel. Therefore, climate change increases the hope of using the Arctic route, which is a promising region for navigation as it potentially halves the distance between the Far East and Europe and could possible save approximately a greater percentage of commercial voyages through Suez or Panama Canal.[6] A technical analysis of the reasons underlying opposing geographical claims in that region would reveal that Arctic oil and gas resources play a modest role among significant economic incentives similar to other maritime claims of the world. The Arctic is bordered by countries that consume a vast amount of hydrocarbons, especially the U.S., Russia, Norway and Canada, who would certainly benefit from a secure and proximate source of natural resources hidden beneath troubled waters.[7] The majority of the exploitable Arctic reserves made available due to climate change lie within the national jurisdictions of a few States which renders them as exclusive economic rights.[8] Other parts are currently under the regime of MOUs but joint development is just a theory. The hypothesis drawn from the economic incentive factor can pave a concrete understanding of the

---

[4] *Supra* note 1, *See* also, Peter Ehlers, "Effects of Climate Change on Maritime Transportation" in *Impacts of Climate Change on the Maritime Industry*, The proceedings of the Conference on Impacts of Climate Change on the Maritime Industry, 2–4 June 2008 (Sweden), Neil Bellefontaine and Olof Linden (eds.), 2009, World Maritime University Publications, p. 52, where it is stated that the NSR leads from Northern Europe around North Cape and along Siberia through the Bering Strait to Asia and shorten 11,000–12,000 nautical mile sea routes from Europe to Asia by 4,000 miles.

[5] *Ibid*, where it is stated that the NWP is a sea route that connects the Atlantic and Pacific Oceans through the Canadian Arctic Archipelago and ship routes via the NWP from Europe to Japan, China and other Eastern destinations would be 2,500 miles shorter.

[6] Nathan D. Mulherin, *The Northern Sea Route: its development and evolving state of operations in the 1990s/Nathan D. Mulherin; prepared for U.S. Army Engineer District, Alaska*, 1996, U.S. Army Corps of Engineers, Cold Regions Research & Engineering Laboratory.

[7] Janelle Kennedy, Arthur J. Hanson and Jack Mathias, "Ocean Governance in the Arctic: A Canadian Perspective" in *Securing the Oceans: Essays on Ocean Governance-Global and Regional Perspectives*, Chua Thia-Eng, Gunnar Kullenberg and Danilo Bonga (eds.), January 2008, Published by GEF, UNDP and IMO in association with the Nippon Foundation, p. 635.

[8] *Ibid.*

reasons why international navigation may witness a sharp increase in the near future. The driving force of economic profit is certainly a justification for the Arctic States to place contradicting claims in the troubled waters.

5.2 Arctic Ocean Governance

Oceans are perceived as a transportation conduit for international trade that affects sustainable development everywhere.[9] *Prima facie* there exist two categories of trade-related activity that have a bearing upon the ocean namely, the use of ocean as a transportation network and the use of ocean production processes.[10] Either ways, when maritime transportation inaugurates in a virgin territory worth a magnitude of commercial incentives, it develops radically and takes its toll on the environment itself. From intentional or accidental "oil pollution"[11] to threats of "invasive species",[12] the pristine environment becomes susceptible to any type of ship-source pollution. Hence, strategies for different forms of ocean governance is formulated which have either direct or indirect impact on trade. These range from the use of direct trade restrictions as a mechanism for impacting the behavior of States in protecting ocean resources, to other regulatory processes, commonly termed as indirect trade restrictions but which have an influence on trade itself.[13] Nevertheless, in the context of pragmatic international trade, the WTO and GATT have laid down rules where States are not allowed to impose discriminatory trade restrictions, the exception of which is subject to environmental considerations.[14] To what extent can this "exception" be applicable depends on the vulnerability of

---

[9] *Supra* note 1.

[10] *Ibid.*, p. 160, para 1, where it is indicated that "[t]he oceans have played both of these roles in human commerce for thousands of years. But the impact of both kinds of commerce on ocean ecosystem has increased exponentially in the past century".

[11] *Supra* note 5 in Chap. 1, *See* Article 1(4), where it is expressly stated that, "…the introduction by man, directly or indirectly, of substances or energy into the marine environment, including estuaries, which results or is likely to result in such deleterious effects as harm to living resources and marine life, hazards to human health, hindrance to marine activities, including fishing and other legitimate uses of the sea, impairment of quality for use of sea and water and reduction of amenities".

[12] Michael Elliot, "Biological pollutants and biological pollution-an increasing cause for concern", *Marine Pollution Bulletin*, 2003, volume 46:1, Publications expediting incorporation, p. 275, whereby it has been explained that biological pollution as opposed to any other ship source pollution has been transparently taken to comprehend the pollution emanating from organisms i.e. nutrients or organic matter, and even pollution affecting biological organisms. The author furthers the interpretation as the effects of introduced, invasive species sufficient to disturb an individual, a population or a community; including the production of adverse economic consequences.

[13] *Ibid*, p. 170.

[14] *Ibid.*, p. 171.

the zones. Arctic ocean governance in this regard needs further clarification. This can be estimated by observing the *status quo* legal regime of regions possessing equal or approximately similar characteristics and a comparative study between the both.

5.2.1 Antarctic Treaty as a Model Regime

Currently the Arctic region consists of an array of specific regimes with very narrow issue areas.[15] The regional regime suffers from lack of co-ordination in structuring protected areas beyond national waters while the legal regime suffers from unenforceability. That global and regional co-operation is necessary to protect and preserve the marine environment, is already recognized in Article 197 of UNCLOS. The Arctic States have failed to implement any pragmatic regional treaty or common integrated policy with respect to protected areas. The AEPS and later the Arctic Council are recent developments which have been successful so far in pinpointing the need for designation of MPAs within and beyond national jurisdiction. Majority of the Council members linger with *ad hoc projects* with no clear agenda.[16] The Antarctic treaty regime is observed by scholars as a framework that should be used in the Arctic as well owing to the fact that the South Pole shares a similar environmental stress factor with its northern counterpart, and is governed by a comprehensive environmental protection treaty i.e. *modus operandi.* Hence, the Arctic region seems incomplete when compared with the far-reaching "zero-discharge" policy regulating the Antarctic.[17] In brief, the Arctic legal regime is less comprehensive when compared with the treaty-based regime that regulates the Antarctic, a region with a similar environment. As such, the point of analysis is whether the Arctic is in need of a new legal regime, and whether it should be modeled after the Antarctic-treaty. Analysis of the Arctic *status quo* reveals that UNCLOS does not explicitly provide any reference to it even if the "Canadian Clause" i.e. Article 234 is interpreted and stretched far to fit its conditions. This sense of deprivation as regards the Arctic region mounts up when investigating the provisions of MARPOL 73/78 which gives no reference to the "numbers and figures" of pollutant discharges in the Arctic. Whether the Antarctic-treaty should be idolized by the Arctic Council is a question the answer to which remains in the analysis of future commercial implications of the Arctic. Although the Antarctic-treaty "zero discharge" policy seems suitable to safeguard the marine environment, the applicability of this strict ocean governance corresponding to sustainable development needs to be determined by placing it opposite to the notion of progressive international trade.

[15] *Supra* note 42 in Chap. 1, p. 420.

[16] *Supra* note 5 in Chap. 4, pp. 131–170.

[17] *Ibid.*

5.2.2 *Sustainable Development Versus Commercial Implication*

Sustainability is complemented by the "security concept" and "security" itself is a spectrum that includes social, environmental and other aspects.[18] In the Arctic, security will include matters related to sovereign control, rights of use and movements of people and goods by sea.[19] Once sovereign control is gained, it will be inevitable for the States to focus on sustainable development so that the resources are not exhausted.[20] Sustainable development can be achieved by keeping the Arctic Ocean free from maritime contamination and "zero-discharge" policy with a liability facet, apparently, is an idea which seems to ensure that. "Zero-discharge" in a technical manner, can be compared to "absolute liability" insofar as the duty of the Flag State is to keep the discharge of pollutants or other wastes to the level of "zero" and in case of failure the defendant will not be given the benefit of the doubt i.e. negligence. The reason behind supporting this policy lies in the fact that the Arctic has not yet been developed for international navigation and in dealing with the pollutants appropriately, since it lacks a number of legal issues and establishments corresponding to protection of the marine environment.[21] The positive aspect of it is that a legal regime complementing "zero-discharge" policy parallel with the Antarctic-treaty will boost the concept of sustainable development, or so the environmentalist extremists would have the world believe. This idea in general, is not compatible with the term "navigation" because it conflicts with the smooth operation of the vessel by restricting a minimal fraction of discharge and emission.

Although the international dictators of environmental policies suggest strict protected MPAs at a regional level for the Arctic, such notion in terms of the Arctic route which halves the distance between Far East and Europe cannot be implemented due to objections which will be raised by the international shipping community who seek commercial interest by being thrifty with time when using

---

[18] G. Kullenberg, "Approaches to addressing the problems of pollution of the marine environment", *Ocean and Coastal Management*, Volume 42, Number 12, December 1999, pp. 999–1018; *See* also, Marc. A. Levy, "Is the Environment a National Security Issue?", *International Security*, Vol. 20, No. 2, Autumn 1995, MIT Press, pp. 35–62.

[19] *Ibid*, p. 644, para 2.

[20] Stephen A. Macko, "Changes in the Arctic Environment" in *Changes in the Arctic Environment and the Law of the Sea*, Myron H. Nordquist, John Norton Moore and Tomas H. Heidar (eds.), 2010, IDC Publishers, Martinus Nijhoff Publishers and VSP, p. 108, where it is stated that "[o]nce the ownership in offshore regions is delineated, exploitation of these resources will occur".

[21] This is observed as a significant issue for the Arctic since well-developed ports is relatively few in number and reception facilities may not exist. Furthermore, there is no port State control or Memorandum of Understanding for this purpose in the Arctic.

those routes.[22] Then again, UNCLOS does not encourage the implementation of "zero-discharge" policy when authorising Coastal States to have effective control over Flag State vessels. "Control" under a logical interpretation means "control from damage" where the general threshold should be the normal quantity of discharge or emission.[23] This normal quantity of discharge does not disparage the theory of sustainable development as opposed to "pollution beyond control". It would be quite unethical for the policy makers to dominate the shipping industry by incorporating "zero-discharge" regulations in the Arctic and suppressing the commercial implication it holds.

It is relevant that under UNCLOS, States have the right to impose special environmental requirements as a condition for access to their Ports and ascertain necessary steps vis-à-vis vessels, regardless of their flag, to prevent any violation of these conditions (Article 2(1)). Port State's prescriptive jurisdiction can be found in Articles 25(2) and 211(3) of UNCLOS which is significant for States to impose requirements and to reject entrance to vessels not complying with these conditions which is a part of customary international law.[24] To what concerns Port State Jurisdiction, and in fact the only provision in UNCLOS that refers to Port State, is Article 218 which confers a right on Port State to investigate, correct and eventually punish violations of International "discharge" standards by foreign ships on the high seas or in any areas under the jurisdiction of another State.[25] This provision is hence, one of the exceptions to the supremacy of Flag State jurisdiction on the high seas which upholds the intention of "sustainable development". On the other hand, in exercising enforcement powers, a set of safeguards are to be

[22] Douglas R. Brubaker and Willy Østreng, "The Northern sea route regime: Exquisite superpower subterfuge?", *Ocean Development and International Law*, Vol. 30, Issue 4, 1999, pp. 299–331, *Note* also, the commercial advantages of the Arctic Sea Route can be estimated from the advantages reaped from the use of the Suez Canal which is considered to be the shortest link between the East and the West due to its unique geographic location; it is an important International navigation canal linking between the Mediterranean sea at Port said and the red sea at Suez.

[23] Reference has been made to Canada's AWPPA which promotes "zero-discharge" policy. Nevertheless, under certain provisions of ASPPR discharge of oil and sewage generated on board the ships is permissible. Hence, "zero-discharge" under AWPPA is coupled with flexibility. But the Author's discussion of "zero-discharge" under this context refers to the idea of a complete restriction of discharge in the Arctic, a presupposition favored by many scholars, *See* also, *Supra* note 2 in Chap. 2, where the main reason for implementation of Canada's AWPPA has been seen as an effort to gain International recognition, an indirect means of being acknowledged for the sovereign claim over the NWP.

[24] The conditions laid down in Articles 2(1) and 211(3) must be given due publicity and communicated to the IMO; Port States can grant or deny entry into their ports is a part of customary international law which has been recognized by ICJ *See also Nicaragua Case* (Nicaragua vs. U.S.), Judgment of 27 June 1986, ICJ Reports, pp. 14–101, para 213, *See* also, David Anderson, "Port States and Environmental Protection" in *International Law and Sustainable Development: Past Achievements and Future Challenges*, Alan Boyle and David Freestone (eds.), 10 May 2001, OUP Oxford, pp. 325–337.

[25] *Supra* note 12 in Chap. 1, p. 205.

observed by the Port State to protect the commercial interest of foreign ships.[26] This balances the commercial interest of those ships against discriminatory yet excessive authority of the Port States.[27] This can also be interpreted as a bar against any stringent policies set up to impede the normal discharge level of a particular vessel. Moreover, IMO instruments, such as MARPOL 73/78 indicates to the regime corresponding to the legislative jurisdiction of Coastal States contained in UNCLOS.[28] This mode of balance as immaculately enshrined in the texts of international conventions cannot be implemented due to the inadequate number of ports in the Arctic region.[29] It seems, establishment of Ports with effective Port State control stripped off discriminatory Regulations consistent with international law can ensure sustainable development. Considering the commercial implications of the Arctic region, sustainable development under no circumstances provides a reason to execute over-rated standards for the shipping industry.

[26] *Supra* note 5 in Chap. 1, Part XII, Section 7, Articles 223–231.

[27] From a general perspective, there must exist a nexus between access requirements and legitimate interests of Port States to ensure environmental protection and should be balanced with the navigation rights of Flag States.

[28] Article 9(2) of MARPOL 73/78 states that, "Nothing in the present Convention shall prejudice the codification and development of the law of the sea by the United Nations Conference on the Law of the Sea […] nor the present or future claims and legal views of any State concerning the law of the sea and the nature and the extent of coastal and flag State jurisdiction".

[29] *Supra* note 9 in Chap. 3, p. 188.

Chapter 6
Corporate Social Responsibility and the Arctic

The concept of "ocean governance" has many dimensions and is therefore, inconclusive in nature.[1] Although there is no agreed-upon definition of ocean governance, one can observe the emergence of certain elements which, when combined, could begin to form the necessary basis for a holistic and inclusive ocean regime. Although "Ocean" has been defined as the holistic nature of the ocean where due recognition must be given to the fact that problems therewith are closely interrelated whereby "Governance" according to scholars can be carried out at an international, regional, national and local level as regards those interrelated problems. Lawrence (1999) has laid out a terse definition of "ocean governance" which is primarily constituted by institutions, formal and informal agreements and behaviors, how resources are used, how the problems and chances are assessed, the actions permitted or prohibited; and the regulations and sanctions that are applied.[2] CSR in this regard is intertwined with the concept of ocean governance in so far as companies integrate social and environmental concerns in their business operations Land in their interaction with their stakeholders on a voluntary basis. "Companies" in this subject matter can be taken to understand the public or private companies functioning as regards shipping business. These are the companies that are to remain accountable to the Arctic society. Every governance regime tries to fulfill a certain objective and when it narrows down to the Arctic, ocean governance and corporate social responsibility have a clear nexus.

---

[1] Rothwell, Donald R., and VanderZwaag, David L., "The Sea Change Towards Principled Oceans Governance", *Towards Principled Oceans Governance: Australian and Canadian Approaches and Challenges*, London: Routledge Press, in press, 2006.

[2] Juda, Lawrence, "Considerations in Developing a Functional approach to the Governance of Large Marine Ecosystems", *Ocean Development and International Law 30*, 1999, pp. 90–91, where the author states that governance is considered a key element in ecosystem management and encompasses the formal and informal arrangements, institutions, and determines how resources and the environment are utilized.

T. Johansson and P. Donner, *The Shipping Industry, Ocean Governance and Environmental Law in the Paradigm Shift,* SpringerBriefs in Law,
DOI 10.1007/978-3-319-12541-1_6

The objective here is environmental protection and the concept of environment is intrinsically linked to society because damage to the environment of the Arctic will eventually have repercussions on the Arctic indigenous inhabitants. Then again, the Arctic Ocean represents a unique socio-cultural entity and in order to regulate this entity, there is a need of a strategic body of law which embraces the core concept of corporate-socio yet an ethical responsibility. According to some authors, the regulatory regime does not stand alone and is connected with other corporate pillars and finally some have drawn the same conclusion that the regulatory regime of human-social elements are associated with ocean governance.

When it comes to theories of CSR, a significant notion that needs to be set is the notion of stakeholder and its association with the specific business in question. Although the debate on the relationships between business and society, and the implied responsibilities, has been on going for decades, there is still no consensus on a commonly accepted definition of CSR.[3] The concept of CSR by itself is sporadically put in relation to other concepts i.e. Corporate Social Responsiveness or Corporate Performance by respective academics.[4] Either ways, an important characteristics of CSR is the idea that business is accountable to various stakeholders who can be identified and have a claim, either legally mentioned or morally expected, on the business

[3] Carroll, A.B., "The pyramid of corporate social responsibility: toward the moral management of organizational stakeholders", *Business Horizons*, 1991, Vol. 34, Iss. No. 4, pp. 39–48. *See* also Jones, T.M., *Instrumental stakeholder theory: a synthesis of ethics*, The Academy of Management Review, Vol. 20, Iss. No. 2.

[4] Kakabadse, Nada K., Rozuel, C and Lee-Davies, L., "Corporate Social Responsibility and Stakeholder Approach: A Conceptual Review", *International Journal Business governance and Ethics*, Vol. 1, No. 4, 2005, p. 280. *N.B.* Table 1, p. 281 which refers to various authors and their definition as regards CSR and business. *See* Bowen, H.R., Social Responsibilities of the Businessman, New York: Harper and Row, 1953, where the author has stated that CSR refers to the obligations of businessmen to pursue those policies, to make those decisions, or to follow those lines of action which are desirable in terms of the objectives and values of our society. A transparent link between social responsibility and utilization of resources has been laid down by Fridman M., "Capitalism and Freedom", Chicago: University of Chicago Press, 1962, where it is highlighted that there is one and only one social responsibility of business—to use its resources and engage in activities designed to increase its profits so long as it stays within the rules of the game, which is to say, engages in open and free competition without deception or fraud. Then again, the relation of business and economics has been stipulated by Carroll, A.B., "A three-dimensional conceptual model of corporate performance", *Academy of Management Review*, Vol. 4, No. 4, 1979, pp. 497–505. This has reflected on the social responsibility of business that encompasses the economic, legal, ethical and discretionary expectations that society has of organisations at a given point in time. From a more Organization point of view, the World Business Council for Sustainable Development (2003) has defined 'CSR' as a business commitment to contribute to sustainable economic development working with employees, their families, the local community, and society at large to improve their quality of life. The hypothesis is that contribution to sustainable economic development in the Arctic would certainly refer to the protection of the marine environment via co-operation because challenges faced by and in the Arctic have increased significantly and have become more visible over the years. The fact that needs to be understood is that there is a need to strike a balance between increased commercialization in the Arctic and the environmental protection regime because the Arctic inhabitants are greatly dependant on the ocean for their sustenance. Greater knowledge has been accumulated about the impact of globalisation and climate change, as well as of the advantages they bring for the Arctic region, the local communities,

activities that affect them.[5] CSR procedures in the shipping industry do not have a well-monitored history when compared to the land based industrial sector. The shipping industry is often termed as more problematic than the land based industries since they operate in different nations and many nations have different laws concerning various environmental regulations. Environmental policies in the shipping industry are usually founded on several standards and certifications. It is observed that the shipping companies have several different certifications especially regarding environment and safety issues. The International Organization for Standardizations ISO 14000 standard series is probably the most well known environmental standard for companies. The purposes of "ISO 14000"[6] series is to assist companies to reduce their processes' and production's impact on the environment and help them in becoming more responsible to the respective stakeholders. The ISO 14001:2004 and ISO 14004:2004 is intrinsically an environmental management system and its requirements and guidelines. The ISO 14000 standards were created after United Nations conference on environment and development in Rio de Janeiro in 1992. The first versions of the standards were made in 1993 to support sustainable development. This is a significant tool for the shipping companies operating in the Arctic since ISO 14001 has been adopted as a

Footnote 4 (continued)
the people of the region, the environment and nature. The Arctic stands on the threshold of major changes, which will not only have an impact on nature, the environment and the climate at global, regional and local level but will also, in the first instance, affect Arctic societies and the Arctic people in a number of important areas. On the one hand, there is the prospect of the loss of traditional ways of life, patterns of settlement and ways of making a living, as well as of natural phenomena and aspects of the environment. On the other hand, there are the innovative aspects of new business and development opportunities done via traversing the waters to exploit oil and mineral resources. It is important to continue to have and to develop processes that take into account the unique Arctic environment and involve local populations, so that decisions are consistent with their needs and desire for self-determination and are also compatible with environmental conservation and with sustainable development in the region. CSR Europé (2003) has added their own definition to this concept where CSR is seen as the way in which a company manages and improves its social and environmental impact to generate value for both its shareholders and its stakeholders by innovating its strategy, organisation and operations. On the other hand, Novethic (2003) has provided a more in-depth definition of CSR and the relations that it has with other dimensions. In short, linked to the application by corporations of the sustainable development principle, the concept of CSR integrates three dimensions: an economic dimension (efficiency, profitability), a social dimension (social responsibility) and an environmental dimension (environmental responsibility). To respect these principles, corporations must pay more attention to all the stakeholders […] which inform on the expectations of civil society and the business environment.

[5] *See* Frederick, W.C., *Theories of corporate social performance*, in S.P. Sethi and M.C. Falbe (Eds.) Business and Society: Dimensions of Conflict and Cooperation, Lexington, MA: Lexington Books/D.C. Heath, 1987. pp. 142–161. *See also* Mitnick, B.M., *Systematics and CSR: the theory and processes of normative referencing*, Business and Society, 1995. Vol. 34, Iss. No. 1.

[6] *N.B.*, although the ISO 14000 standards are designed to be mutually supportive, they can also be used independently of each other to achieve environmental goals. The whole ISO 14000 family of standards provides management tools for organizations to manage their environmental aspects and assess their environmental performance.

national standard by more than half of the 160 national members of ISO and its use is encouraged by governments around the world. A reasonable assumption that can be gathered from this analysis is that, there is no lack in international instruments and CSR environmental tools for the Arctic. The regulations promulgated by IMO and national legislation of the States bordering the Arctic seem to be more than sufficient.

What is left to analyse is the peer pressure drawn from these international and national legal regimes which is eventually the shipping industry in a state of quandary. Before understanding the "balance", it is important to note if there can be a balance. A significant aspect of CSR includes "safety" and the notion of safety is embedded therein because the popular belief is that a safe crew will be able to avoid maritime incidents which will ultimately wreak havoc on the pristine waters.[7] Then again, safety is a multidimensional term which can be used in terms of protection of the marine environment. But protection of the marine environment does not only imply the efforts to protect the marine environment of the Arctic. It also stipulates that the shipping industry is accountable for every fraction of an action in its navigation and environmental management of the ship used for commercial purposes. Accountable in this sense would mean an action towards a certain group. They are the stakeholders of the Arctic.

6.1 "Stakeholder Theory" Before the Balance

Although the "stakeholder theory" is an issue that has been broadly researched over for more than two decades, it is said to ideally embrace the notion that is quite the opposite to the "shareholder model" where every individual entity

[7] *N.B.*, author Laura Arat (2011) has explained that there are several national and international regulations for maritime safety. The most well known and most used of them is the International Maritime Organization's International Safety Management Code (ISM Code), which is part of the international convention for the safety of life at sea (SOLAS). The IMO issued ISM Code in order to guarantee the safe operation of vessels and also to prevent pollution of marine environment. The ISM Code system was founded in 1987 from the resolution A.596(15), in which IMO's Assembly requested that Maritime Safety Committee urgently had to develop recommendations for the management of Ro–Ro passenger vessels to guarantee their safe operation. In October 1989 resolution A.647(16), IMO guidelines for the management of the safe operations of ships and of pollution prevention was developed. These following resolutions created the basics of the present day ISM Code. However it was still a guideline. In November 1993, the guidelines were replaced by the resolution A.741(18), which is also the current ISM Code. The code was reinforced in December 2000 and the ISM Code came into effect worldwide in several phases between 1998 and 2002. Since IMO's ISM Code is not anymore just a guideline for safe shipping, but they are ratified by several governments, it is no longer actually a CSR procedure, since it is not based on voluntarism. However, shipping industry is more regulated industry than the other industries, so IMO and ILO standards and regulations already set a certain level of responsible business practices to shipping industry. Although it does not cover all the values and notions that CSR seeks to achieve.

connected with the business benefits therein.[8] It has been held that only the shareholders in the "shareholder model" can have a legal claim on the purpose of the firm they own and operate.[9] Before trying to set a balance in the *status quo* Arctic regime, it is important to observe some key "stakeholder theory" statements as propounded by Jones and Wicks (1999). These statements are indicative of the fact that the firm has relationships with the stakeholder groups and the processes and outcomes associated with these relationships are based on "interests" whereby those "interests" of all legitimate stakeholders carry a value.[10] Then again Jones et.al. (2000) have surmised that in order to perform well, managers of the business will need to pay attention to a wide array of stakeholders which entails obligations to stakeholders which include, but extend beyond, shareholders.[11] Moreover, Simmons (2004) has provided a more business driven perspective relating to the "stakeholder theory" where organisations have stakeholder groups that affect and are affected by them; these interactions impact on specific stakeholders and the organisation; and perspectives of salient stakeholders affect the viability of strategic options.[12]

Although these reflect theories which specifically deal with elements and situations of business administration, nevertheless, when it narrows down to striking a balance between the increased commercialization of the shipping industry and the laws and concepts of ocean governance, it is significant to acknowledge the entities that affect and are affected by them. In the context of the Arctic and in an effort to fill in the equation with the appropriate entity, the shareholders would be denominated as the big players and the small players of the shipping industry who have a direct interest in the oil and gas explorations in the Arctic. The managers of the business would relate to the shipping companies that are owned by them and the stakeholders would refer to the entities that are ultimately affected by those business explorations. Then again, business in the Arctic centres around exploitation of resources and cannot be termed as negative. When this is integrated with the "stakeholder theory" then it implies that the business in question should contribute to the development or benefit the stakeholders who are directly or indirectly affected. Although the balance is between the shareholders (managers of the business) and the international law that governs the environmental impact of the business, the stakeholders are a part of the "business chain" where such business conducted is

---

[8] *Supra* note 5, p. 289.

[9] Halal, W.E., "Corporate community: a theory of the firm uniting profitability and responsibility", *Strategy and Leadership*, 2000, Vol. 28, Iss. No. 2. *See* also Emiliani, M.L., *A mathematical logic approach to the shareholder vs stakeholder debate*, Management Decision, 2001, Vol. 39, No. 8, pp. 618–622.

[10] Jones, T. and Wicks, A., "Convergent stakeholder theory", *Academy of Management Review*, 1999, Vol. 24, pp. 206–221.

[11] Jones, T.M., Wicks, A.C. and Freeman, R.E., *Stakeholder theory: the state of the art*, in N.E. Bowie (Ed.). 2002, The Blackwell Guide to Business Ethics, Oxford: Blackwell.

[12] Simmons, J., "Managing in the post-managerialist era: towards socially responsible corporate governance", 2004, *Management Decision*, Vol. 42, Nos. 3–4, pp. 601–611.

within or around the domicile or jurisdiction of the stakeholders. This can interpreted as a type of "interconnectedness".[13] Hence, the balance must set in a way that every entity at the opposite ends of the equation can proceed with their own actions and set a check and balance for each other in a "win–win" situation.[14] But now it leads to the important question as to who are the "stakeholders" in the "stakeholder theory" of the Arctic.

6.2 Legitimate Stakeholders of the Arctic

The popular media puts the limelight on climate change and the melting of sea ice and the prospect that this might unleash a rush to extract reserves of oil and gas, exploit clandestined mineral resources and launch new industrial fisheries, use the Arctic as a preferred route for commercial shipping, and treat the region as an increasingly popular destination for tourists. According to author Young (2010), developments of this sort are certainly possible.[15] According to a 2008 USGS assessment, the Arctic accounts for about 13 % of the undiscovered oil (90 billion barrels of undiscovered, technically recoverable oil), 30 % of the undiscovered natural gas and 20 % of the undiscovered natural gas liquids (44 billion barrels of technically recoverable natural gas liquids) in the world.[16] The USGS assessment also states that these resources in the North of the Arctic Circle account for about 22 % of undiscovered and technically recoverable resources in the world.[17]

The demands and the needs for energy and other generic resources are ever-increasing as world economy continues to expand. The reserves of some important resources like oil are becoming more difficult to replace which is frustrating the energy sectors. This is forcing the energy and mining companies to scout for new sources and replacements of supply in frontier regions—such as the Arctic that

[13] *Supra* note 10, where the authors have referred to the 1920s and 1930s which first saw the theoretical developments on a more extended role of business towards society, especially the work of Mary Parker Follett. This recognized a sense of "interconnectedness" among various constituents participating in business performance, and that of Chester Barnard in 1938 which stated that corporations were not an end *per se* but a means to serve society.

[14] *Ibid.*, *See also Supra* note 11, where it has been inferred that The ultimate purpose of the stakeholder management theory is to achieve a 'win–win' outcome, most probably in a medium- to long-term perspective, but it does not infer that managers can manipulate stakeholders to reach the most favourable trade-offs (Post et al. 2002). Stakeholder management aims to allocate resources adequately given the outcome of prospective scenarios; it also aims to repair previous damages traditionally left to society's kind-heartedness, which are referred to as "externalities".

[15] Young, Oran R., "Arctic Governance—Pathways to the future", *Arctic Review on Law and Politics*, vol. 1, 2/2010 p. 164–185. ISSN 1891-6252.

[16] Official website of U.S. Geological Survey, *90 Billion Barrels of Oil and 1,670 Trillion Cubic Feet of Natural Gas Assessed in the Arctic*, available at; http://www.usgs.gov/newsroom/article.asp?ID=1980&from=rss_home#.U9SsiIB_u24 (date accessed 27 July 2014).

[17] *Ibid.*

until recently had attracted minimal interest. With this increased interest circling around the Arctic, nations far and near have risen to the occasion to occupy a place in the "stakeholder table". However, debates as to the membership in the circle of those possessing a legitimate claim to recognition as stakeholders in Arctic affairs have not been prominent in recent years.[18] In the aftermath of the state change of the 1980s/1990s and as an outgrowth of considerable debate, the eight Arctic states operating as the members of the Arctic Council, have dominated the scene.[19] Jurisdictional problems in those in terms of the NWP and the NSR have also been reasons as to why they have dominated the scene and acted as leading authorities. Moreover, the Council has accepted a number of indigenous peoples organizations as "permanent participants" in its activities, a notable precedent with inferences extending far beyond the Arctic.[20] Other non-state actors have had to settle for observer status.

At the invitation of the Danish Minister for Foreign Affairs and the Premier of Greenland, representatives of the five coastal States bordering on the Arctic Ocean—Canada, Denmark, Norway, the Russian Federation and the United States of America rendezvoused at the political level on 28 May 2008 in Ilulissat, Greenland, to hold discussions.[21] This meeting was followed by the adoption of the Ilulissat Declaration. In the Ilulissat Declaration, the five Arctic States, i.e. Canada, Denmark/Greenland, Norway, Russia, and the US—presented themselves as coastal states occupying a "unique position" to address issues relating to the Arctic Ocean, broke ranks with the Arctic Council, and asserted a new formula for the identification of stakeholders regarding an important component of the Arctic agenda.[22] This has been marked by criticism from the international community and has been marked as a division of labor between the Arctic Council and other venues for addressing various Arctic subject-matters.[23] A significant controversial

[18] *Supra* note 15.

[19] *Ibid.*

[20] *Ibid., See also* official homepage of the Arctic Council, available at; http://www.arctic-council.org/index.php/en/about-us/permanent-participants/123-resources/about/permanent-participants (date accessed 27 July 2014), where the reasons of such participation has been clarified with the statement that "[o]ut of a total of 4 million inhabitants of the Arctic, approximately 500,000 belong to indigenous peoples. Indigenous peoples' organizations have been granted Permanent Participants status in the Arctic Council. The Permanent Participants have full consultation rights in connection with the Council's negotiations and decisions. The Permanent Participants represent a unique feature of the Arctic Council, and they make valuable contributions to its activities in all areas."

[21] The Ilulissat Declaration, Arctic Ocean Conference, Ilulissat, Greenland. 27–29 May 2008. Available at; http://www.oceanlaw.org/downloads/arctic/Ilulissat_Declaration.pdf (date accessed 27 July 2014).

[22] *Ibid.*

[23] *Ibid.*, the author has also reflected on the effects of such declaration which has had broader impacts as well. The declaration, according to the author, asserts that the coastal states are committed to the framework provided by the law of the sea and that they will behave as responsible stewards regarding the ecosystems of the Arctic Ocean, others are not entirely convinced by

aspect of the background of the Ilulissat Declaration notes that this form of Declaration is based on what the five Arctic States consider as a part of "sovereignty", sovereign rights and jurisdiction in large areas of the Arctic Ocean. But the concept of sovereignty under maritime law is a paradox and the jurisdiction aspect of the Arctic Ocean is shrouded by ambiguity. Although the five Arctic States have based their consensus on relevant "protection of environment" articles of UNCLOS, they have, nevertheless, defied the "stakeholder" theory which takes into consideration every entity that is directly or indirectly affected by the business in question. This would, to a greater extent, exclude the "permanent participants" which is represented by the local native indigenous of the Arctic.[24]

The breadth of the Arctic Council's reach can be seen in its sub-bodies and has an Indigenous Peoples Secretariat that compiles information and develops recommendations for the Council. Then again, other cooperative and consultative bodies are also active comprising of the Nordic Council which includes several EU member states (Denmark, Finland, and Sweden), Norway, Iceland, and autonomous territories of Greenland (the Faroe Islands and Aland Islands) and the Barents Euro-Arctic Council provides a mechanism for cooperation in the Barents region.[25] The Ilulissat Declaration with its "grounds for formation" frustrates the objective with which the Arctic Council set to achieve. This unilateral move of the five Arctic States has influenced other non-Arctic states and organizations to assert claims and place demands to be treated as legitimate stakeholders with regard to

---

Footnote 23 (continued)
this argument. It is for fact that both the Arctic Ocean proper and the seabed beyond the outer limits of the jurisdiction of coastal states are international commons. All the signatories to the UNCLOS have a legitimate interest in what happens in and to these commons. Then again, recognizing talked about expansive boundaries of coastal state jurisdiction over the seabed under the provisions of UNCLOS, Article 76 will of necessity reduce the scope of the commons, a development that has obvious consequences for the interests of others. Some have suggested that the coastal states would be justified in adopting the view that the Arctic Ocean is an enclosed or semi-enclosed sea under the terms of UNCLOS, Part IX, a position that would accord them authority to manage this region. The author believes that it is unlikely that other states will acquiesce in such a claim. In any case, UNCLOS Article 123 calls on coastal states to co-operate with "other interested States" in managing activities occurring in enclosed or semi-enclosed seas. *See also Supra* note 5 in Chap. 1.

24 West, Mary B., "Arctic Warming: Environmental, Environmental and Security Implications", *Vanderbilt Journal of Transitional Law*, 2012, Vol. 42:1081, where the author has stated that global warming as a result of climate change will have direct effects on the environment and the health and well-being of Arctic residents. The slow pace of natural recovery in the Arctic makes it critical to address such effects as quickly as possible. In addition, the potential impacts of intensified Arctic warming on the Earth's oceans and climate in general make the need to address environmental change in the Arctic even more critical. In creating the Arctic Environmental Protection Strategy (AEPS) in the early 1990s to address concerns related to transboundary pollution, the international community recognized the importance of cooperative action to address Arctic environmental security.

25 Norden, About Nordic Co-operation, available at; http://www.norden.org/en (date accessed 27 July 2014), *See also* Csonka, Yvon and Schweitzer, Peter, *Societies and Cultures: Change and Persistence*, in Arctic Human Development Report 45, 47, 53 (Niels Einarsson et al. eds., 2004).

Arctic issues. The European Parliament has passed a resolution in 2008 specifying the interests of the EU in the Arctic and staking out a claim to be accepted as a legitimate stakeholder in addressing Arctic issues.[26] The claim of Arctic stakeholdership by individual States has ignited Asian counterparts to slowly pave their own way which is evident from the action of China which has proceeded with a number of steps to assert its right to be treated as an Arctic stakeholder, especially with regard to matters relating to the Arctic Ocean.[27] Japan is developing a generic position, albeit in the low-key manner characteristic of Japanese diplomacy.[28]

The ultimate question was posed by author Young (2010) as to the implications of these developments with regard to Arctic governance. What can be added to this inevitable question is the *status quo* existence of the "stakeholder theory". Author Young (2010) implicitly states that there no "point-of-return" to the previous informal consensus regarding the role of the eight Arctic States (in the Arctic council) as the primary stakeholders in Arctic affairs.[29] The author furthermore stresses that whatever may be the merits of the situation in normative terms, the current state change is strengthening the links between Arctic affairs and global processes, an occurrence that is shifting the geopolitical underpinnings of Arctic issues.[30] The

---

[26] Official Website of the European Union Parliament, available at; http://www.europarl.europa.eu/sides/getDoc.do?type=TA&reference=P6-TA-2008-0474&language=EN (date accessed 27 July 2014). *N.B.* One facet highlighted in the resolution is the fact that the Arctic region is currently not guided by any specific formulated multilateral norms and regulations, as it was never expected to become a navigable waterway or an area of commercial exploitation. The EU Parliament in its declaration (Article 1 of the respective Resolution) has stated that it is deeply concerned at the effects of climate change on the sustainability of the lives of the indigenous peoples in the region, in terms of both the general environment (melting icecap and permafrost, rising sea levels and flooding) and the natural habitat (the retreating icecap poses problems for polar bears' feeding habits), and underlines that any international decisions relating to these issues must fully involve and take account of all peoples and nations of the Arctic. Moreover, the Resolution suggests that suggests that the Commission should be prepared to pursue the opening of international negotiations designed to lead to the adoption of an international treaty for the protection of the Arctic, having as its inspiration the Antarctic Treaty, as supplemented by the Madrid Protocol signed in 1991, but respecting the fundamental difference represented by the populated nature of the Arctic and the consequent rights and needs of the people and nations of the Arctic region; believes, however, that as a minimum starting-point such a treaty could at least cover the unpopulated and unclaimed area at the centre of the Arctic Ocean.

[27] Jakobsson, Linda, *China Prepares for an Ice-free Arctic*, 2010, SIPRI Insights on peace and Security, No. 2010/2. From a rather analytical perspective the author has reflected on China's economy which is reliant on foreign trade and the fact that there are substantial commercial implications if shipping routes are shortened during the summer months each year. Nearly half of China's gross domestic product (GDP) is thought to be dependent on shipping. Chinese Officials have stressed that China's Arctic research activities remain primarily focused on the climatic and environmental consequences of the ice-melting in the Arctic. However, in recent years Chinese officials have started to also assess the commercial, political and security implications for China of a seasonally ice-free Arctic region.

[28] *Supra* note 15.

[29] *Ibid.*

[30] *Ibid.*

secret agenda of a short sea route to save fuel oil is now out in the open. Although the number of non-observer states to have control over the Arctic, these attempts may prove to be futile in the long run.

The paradigm shift of unique control has already begun and is very evident from the Ilulissat Declaration. That is why author Young (2010) strongly believes that if the current forecasts regarding the region's natural resources prove to be correct, the Arctic will be seen by players like China and the EU as too important to be left to the "Arctic 5" with regard to matters of governance.[31] The other three members—who are quite opposed to the unilateral movement of the "Arctic 5" are unlikely to come to the defense of the coastal states in this realm. This conflict within the inner circle of the Arctic Council will lead to a variety of non-state actors to seek opportunities for advancing their own interests in this unsettled situation. The problems of the NWP and the NSR will add incentives to those non-state actors to declare themselves as stakeholders which will ultimately frustrate the so-called business jargon of "stakeholder theory". Then again, the "stakeholder theory" does so much as to include distant entities that could be or might be affected by the business in an indirect fashion although they might not be immediately related to the impacts caused by the business i.e. oil and gas explorations in the Arctic. This under the context of the current political issues subsisting in the Arctic Council may not be the best and ideal way to determine the number of stakeholders.[32] The Arctic at this point does not need any off-distant or non-state stakeholders except for the States that are directly linked to the Arctic. These Arctic States should promote the CSR aspect of shipping and should compel Flag States and respective shipping companies to commence CSR reportings in black and white. The Arctic States should not deter from the objective of protecting the pristine environment and promotion of ocean governance. The proposition as laid down by author Marffy (2004) could be supported at this stage which states that ocean governance comprises of four pillars i.e. legal pillar, the political pillar, the institutional pillar and the capacity-building pillar and that the legal pillar should be positioned at the top of the hierarchy.[33] If the juxtaposition of the legal pillar is

---

[31] *Ibid.*

[32] *Ibid.*, where the author has vividly stated that these developments regarding stakeholders could prove disruptive for the Arctic Council, a body that owes its existence to the decoupling state change of the 1980s and 1990s. The author continues to state that, because the Council lacks the authority to operate as a regulatory body, some stakeholders—including actors located in the Arctic—may see little reason to be distressed by the impacts of such disruptive forces. Still, we can expect a period in which a fair amount of time and energy will be expended on debates about the identity of the relevant stakeholders and about the character of the Arctic Council, in contrast to focusing on the substantive concerns giving rise to a demand for governance in the Arctic. According to the author, this is not necessarily a bad thing. But to the extent that developments like climate change produce a need for prompt.

[33] Marffy, Annick de, *Ocean Governance: A Process in the Right Direction for the Effective Management of the Oceans*, 18 Ocean Y.B. 162, 2004, Division for Ocean Affairs and the Law of the Sea, Office of Legal Affairs, United Nations, New York. This rather new concept of ocean

to be at the top of the hierarchy, then it will be inevitable to set the balance in terms of "hard law" and "soft law". Because the concepts of CSR and stakeholder are interrelated, the two approaches need further examination from a broader perspective and put in relation in theory and in practice, in order to reposition their underlying principles into a broader assessment of the relationships between

Footnote 33 (continued)
governance has acquired only recently a more meaningful content. Although still a process in need of further progress on the road leading to its effective implementation, it has certainly moved a step forward. The first thinking about the necessity of establishing an efficient governance of the oceans, taking into account all the synergies existing between activities in the oceans, started during the meeting of *Pacem in Maribus XIX* in 1991. No definition of the concept has been offered so far. However, based on the principles of holistic and inclusive approach set out in the Convention, which states in its preamble that "...the problems of ocean space are closely interrelated and need to be considered as a whole..." four main pillars could be seen as constituting the foundation to effective governance. They comprise of the legal pillar, the political pillar, the institutional pillar and the capacity-building pillar. Marffy has succinctly focused only on the international aspect of ocean governance and on the understanding that it is essential that those four pillars also be implemented *mutatis mutandis* at the national level. The interdependence of the regulatory regime as regards ocean governance with other pillars is a significant hypothesis of this literature. On a parallel level Yankov (2004) has focused on the regulatory-genre in terms of protection and development of the marine environment and has highlighted on the nexus between Ocean governance and Corporate Social Responsibility. One of these is the integrated concept of protection and sustainable development of the marine environment and ocean governance. This brings together in an inclusive and interdependent manner two most important aspects of sound management of the seas and oceans, namely, protection and conservation of the marine environment, on the one hand, and economic and social development of coastal and marine areas, on the other. This is something that could be applicable in terms of the Arctic. The promotion of this concept, according to Yankov, was one of the guiding objectives in Elizabeth Mann Borgese's activities. Then again, Borgese's understanding of ocean governance implied that the international legal order to be established through UNCLOS should, as stipulated in the Preamble of the Convention, be in conformity with the principles of the new international economic order. The concept of the integration of environment and sustainable development of the marine environment is the result of the interface between UNCLOS and UNCED, in particular Agenda 21's Chapter 17, the Rio Declaration on Environment and Development and the conventions, agreements, action plans, and documents from conferences and other international instruments that followed UNCED. The elucidation of the objectives and mechanism of this interface required at the outset a brief examination of the evolving policy of protection and preservation of the marine environment under the influence of UNCLOS and special conventions and agreements relating thereto, on the one hand, and the emerging concept of sustainable development, as a follow-up to the Report of the World Commission on Environment and Development and UNCED, on the other. Elisabeth Mann Borgese made a valuable contribution to this problem area through several IOI projects and most of the *Pacem in Maribus* Conferences since 1987–1988. In addition, to complete and augment the examination of the interrelationship between ocean governance and sustainable development, the input of the Johannesburg Summit 2002 and in particular, the Johannesburg Declaration on Sustainable Development, should be noted. The emergence of ocean governance had several catalysts. There was a growing recognition in developed nations of coastal degradation due to inappropriate development and poor planning. Globally, it had become evident that there was a need to conserve and optimally utilise the resources, protect and preserve the marine environment and coordinate activities within the UN. The 1972 UN Stockholm conference on the human environment reflected the first effort by the international community to deal with environmental problems on a comprehensive basis by creating the UNEP. Then again,

business and the Arctic society and environment. It is apparent however, as research tends to demonstrate, that the drivers of the shipping industry, in particular shipping companies, have a significant role to play in enhancing the social and environmental performance of their organisations as regards accountability for the Arctic. If these shipping organisations are to mature towards greater responsibility, the Arctic States should essentially first reflect on their own position, behaviour, jurisdictional issues, conflict of law and expectations.

Footnote 33 (continued)
UNCLOS provided a comprehensive framework for the governance of the oceans. The 1992 UN Conference on Environment and Development by underlying the need for interdependence and integration developed the basis for ocean governance. The 1990s saw the emergence of international guidelines on the concepts through OECD in 1991, World Bank and International Union for the Conservation of Nature and Natural Resources in 1993 and the World Coast Conference report in 1994. When it narrows down to protection of the environment, it seems that ocean governance should comprise of a holistic, inclusive, and accountable order that is in accordance with international law. Chaturvedi (2001) has, on the other hand, articulated a three-point manifesto. The first of these three manifestos has reflected critically on the notion of ocean governance itself, essentially by way of highlighting the often-ignored distinction, and at times the conflict, between the ethics and politics of sustainability. As regards the second purpose, Chaturvedi has examined the extent to which the politico-legal-ethical dimensions of sustainability have found a place in the dominant intergovernmental geopolitical discourses and interstate agreements in both the Antarctic and the Arctic. The author begins with a brief overview of contemporary geopolitical transformations, both discursive and actual, in the Antarctic and the Arctic. Hence, the Antarctic regulatory regime for ocean governance has been brought into discussion. This is followed by an assessment of the current relevance of the existing law of UNCLOS for ocean governance in the polar-regions. The third argument was derived from the prospects of effective ocean governance of the polar seas depend not only on the further evolution of the international UNCLOS to address specific attributes and requirements of the polar regions, but also on the political will and willingness of participants (state as well as non-state) to negotiate and implement circumpolar norms and legally binding agreements; to look beyond the narrow polar constituencies and concerns; and to address broader principles, practices, and problems of ecologically sustainable, equitable, and peaceful ocean governance at local, regional, and global levels. In this regard, Chaturvedi questions the effectiveness of Borgese's framework of "The oceanic circle" which embodies the legal, political, and ethical principles and parameters of ocean governance. Chaturvedi (2001) has, on the other hand, articulated a three-point manifesto. A brief critique is also attempted of the concept of sustainability as has been articulated and defined both by UNCLOS and Agenda 21, and more recently by the IWCO as well as a submission to the Club of Rome by one of its founding members, Elisabeth Mann Borgese. The first of these three manifestos has reflected critically on the notion of ocean governance itself, essentially by way of highlighting the often-ignored distinction, and at times the conflict, between the ethics and politics of sustainability. As regards the second purpose, Chaturvedi has examined the extent to which the politico-legal-ethical dimensions of sustainability have found a place in the dominant intergovernmental geopolitical discourses and interstate agreements in both the Antarctic and the Arctic. The author begins with a brief overview of contemporary geopolitical transformations, both discursive and actual, in the Antarctic and the Arctic. Hence, the Antarctic regulatory regime for ocean governance has been brought into discussion. [It has been emphasized that against the backdrop of easily identifiable patterns of short periods of ruthless exploitation of natural resources, followed by the collapse of these resources, and finally, an extremely slow recovery, the Antarctic Treaty States negotiated the CCAMLR in the 1970s. Concluded in 1982, the process of negotiating CCAMLR was largely dictated by three common concerns of the ATCPs.

6.3 An Inconclusive Theory

CSR is a concept, which lacks a definition in general. It is an umbrella term for a concept of operations that companies use to act responsibly and hold themselves accountable to the people and other entities in a corporate and social manner. This

---

Footnote 33 (continued)

First, conservation of krill might become extremely difficult if ongoing harvesting was not restrained. The second concern was to tackle effectively the sovereignty problem so that the proposed fisheries management regime in the Southern Ocean did not dilute or demolish the legal position of either claimants or non-claimants to the status of claims to Antarctica. Moreover, The CCAMLR is thus the major convention addressing environmental conservation and resource management in the Antarctic. It applies to all marine resources within an area delineated by the mean position of the south polar front, and as such its northern boundary differs from that of other conventions (600S). The principal institutions of CCAMLR are the Commission as the policy-making and regulatory body, and the Scientific Committee as an advisory body and a forum for consultation and cooperation in compilation and exchange of information]. This is followed by an assessment of the current relevance of the existing law of UNCLOS for ocean governance in the polar-regions. The third argument was derived from the prospects of effective ocean governance of the polar seas depend not only on the further evolution of the international Law of the Sea to address specific attributes and requirements of the polar regions, but also on the political will and willingness of participants (state as well as non-state) to negotiate and implement circumpolar norms and legally binding agreements; to look beyond the narrow polar constituencies and concerns; and to address broader principles, practices, and problems of ecologically sustainable, equitable, and peaceful ocean governance at local, regional, and global levels. In this regard, Chaturvedi questions the effectiveness of Borgese's framework of "The oceanic circle" which embodies the legal, political, and ethical principles and parameters of ocean governance [the author has highlighted on some of the questions that The Oceanic Circle forces us to address, both in general as well as in polar-specific terms, are as follows: How do we realize a paradigm shift from the old geopolitics of mastering the oceans to a new geopolitics of governing the oceans in accordance with the spirit of the Common Heritage of Mankind, with its economic, environmental, and disarmament dimensions? How do we develop a "normative vision of ocean governance and management, cognizant of the transparency of the boundaries both horizontal (as between disciplines, departments, ministries, specialized agencies and programmes) and vertical (as between levels of governance: local, national, regional and global), and committed to the principles of sustainable development, common and comprehensive security, equity, common heritage and participation." And finally, how do we develop an institutional framework of ocean governance that is comprehensive (i.e. reaching from the local level of community through the levels of provincial and national governance to regional and global levels of international organization), consistent (i.e. ensuring that regulations and decision-making processes at all levels of governance must be compatible), and participational (i.e. enabling communities that are normally at the receiving end of central/federal regulations and often marginalized, to have an effective say in the making as well as implementation of such regulations)]. In terms of understanding the chain of logic in the Arctic, due focus can be given on the works of author Rubenstein (2006) who demonstrates the importance of a chain of logic that explains the causal links between ocean governance policies and results. This chain of logic is predicated upon clear accountability for results. Accountability is then a necessary condition for the achievement of ocean governance results during the early years of implementation. Indicators of Accountability are proposed as proxy measures of the early efficacy of the hereby referred to as Collaborative/Ecosystem Oceans Governance. Detailed diagnostics, which are intended to provide an early warning of the ineffectiveness of an oceans governance regime, are developed, based on the Government of Canada's

concept is simultaneously significant for establishing ocean governance in the Arctic. CSR is, nevertheless, built upon voluntary basis, and it seeks to overcome the ethical standards set by law. Many shipping companies have acted responsibly long before the term CSR was introduced into business, however, these companies

---

Footnote 33 (continued)
experience with the Oceans Act (it has been stated that Under the Oceans Act, the Government of Canada developed a new oceans policy based on two principles: an ecosystem-based approach and collaboration among the users of the ocean. It is understood that the objective of the proposed new governance framework was to weave together an "adaptive" oceans governance institution at a geographic scale commensurate with the scale of the estuarine, coastal and marine environments to be "managed" for human use. This framework transcended existing political boundaries defined in constitutional arrangements between the Government of Canada and the Provinces. It was "adaptive" in the sense that it was largely based on trial and error rather than a traditional command-and-control model). The author explains that past oceans management in Canada had suffered from fragmentation, gaps and overlap in jurisdictions and add that, those attempts at oceans governance had primarily been concerned with more traditional ocean uses such as shipping and commercial fisheries. The new approach to oceans governance implicit in the Oceans Act is Integrated Management; an attempt to deal with these past failures. The application of these Indicators of Accountability reveals a problem of "fit," or the extent to which the new oceans governance model is compatible with the traditional accountability conventions of the machinery of the Government of Canada. The application of the Indicators of Accountability reveals four other important lessons learned from the Canadian experience. First, a collaborative process has large transaction- costs in terms of months and years of coordination and participation time required to achieve program coherence within a department, within government and between national, subnational governments and other users of a LOMA. Secondly, before a lead agency or department can effectively collaborate with oceans users outside the department or lead agency, there is a critical need to achieve strong internal collaboration and cooperation within the lead agency or department. These phenomena suggest a CAP. The need for accountability increases with the number of internal and external "collaborators." Paradoxically, the ability to clearly define joint and several accountabilities is inversely related to the number of partners in a collaborative governance regime. Thirdly, there is a critical need to provide a strong basis for accountability in the fundamental oceans policy or enabling legislation. There needs to be clarity around which department or agency is in the lead, as well as the obligations of the "follower" departments or agencies. Furthermore, there must be definitional clarity around the intended meaning of ambiguous terms such as an ecosystem approach, sustainable development, integrated management or the precautionary principle. Fourthly, and perhaps most importantly, there may be a correlation between a lack of accountability and a lack of capacity, competence, commitment to policy coherence and consequence. For example, a lack of a sense of accountability within a given department or agency to support a cross-cutting issue such as an ecosystem approach may undermine a department's capacity to re-assign the scientific resources necessary to develop measures of Marine Environmental Quality. Moreover, author Yen-Chiang Chang (2009) has provided with a critique of the current policy development in relation to oceans governance at the international level. He portrays that the existing mechanisms provide only sectoral governance structures and that there are no clear mechanisms or policy approaches in place to foster cooperation and coordination in a way that could comprehensively and effectively address the conservation of marine ecosystems. Rather, there are various bodies that have specific objectives of their own. At the European level, the current shortcomings of a sector-based approach have been identified, and as a result, a more holistic good ocean governance approach is being encouraged. As the EU Maritime Green Paper states, "Principles of good governance suggest the need for a European maritime policy that embraces all aspects of the oceans and seas." The European Commission further explains that good governance and an integrated approach could

do not represent the entire world fleet. Indeed, if it is CSR that aims to define "what" responsibilities the shipping business ought to fulfil, the stakeholder concept addresses the issues of "whom" the shipping business is or should be accountable to and hence, the interrelation. Fair and ethical as it may be, adopting

---

Footnote 33 (continued)
help to move towards a more overarching strategy, which would more effectively unite the present sectoral policies for maritime activities and environmental policy. Unfortunately, the European maritime policy does not seem to indicate clearly how good ocean governance might be achieved. The current policy proposals in the United Kingdom also do not set out a clear approach as to how good ocean governance can be achieved. In 2002, the U.K. government published. Safeguarding Our Seas, which sets out its vision for "clean, healthy, safe, productive and biologically diverse oceans and seas". In particular, it emphasizes the importance of stakeholder involvement and good fisheries governance. In Scotland, Sea the Opportunity states that one of the key elements of maritime strategy is "promoting good governance." "A Sea Change-A Marine Bill White Paper" referred to the aforesaid proposal and it also emphasizes the importance of "promoting good governance" as an overarching approach to achieving the United Kingdoms sustainable development goal. Unfortunately, none of the aforementioned proposals indicate clearly as to what elements contribute to good governance. The central issue to be considered throughout this article is how best to govern the marine environment. Sustainable development is of primary importance when considering marine environmental protection. The concept of sustainable development, however, has been criticized as being too vague and imprecise in content, and as a result, a clear and workable indicator is essential if progress is to be achieved. "Good governance" is considered to be a positive and constructive element of sustainable development. It is an open decision-making process that should involve public participation, the release of environmental information, and access to environmental justice. While there is a great deal of literature that considers the issue of "good governance," what constitutes the elements of good governance is still the subject of debate. Succinctly, it has been proposed by the author that there are eight elements that appear to be accepted as being the essential elements of good governance, which are, rule of law, participation, transparency, consensus-based decision making, accountability, equitability and inclusiveness, responsiveness, and coherence. These elements of good governance are considered to some extent to overlap. It is apparent that numerous scholarly articles underscore the importance of regional cooperation and maritime regimes for ocean governance across the globe. While some authors have termed the regulatory regime of ocean governance as being 'blunt', some authors argue that the umbrella Convention itself acts as a 'model'. However, most authors believe that the new legal regime for the oceans assumes that the problems of the ocean space are closely interrelated and need to be considered as a whole, where all nations will cooperate in the management of the oceans. More specifically, multiple channels of contact could provide effective policy coordination overcoming some inherent pitfalls of nongovernmental influence on policymakers. On a more general level, policy coordination for the future requires nation-states to act from long foresight and a prescribed system rather than from seeking short-term goals. A fact can be drawn from the scholarly works of ocean governance is that every element should be connected and must be placed in order according to a strategic regulatory regime. Some authors have reflected on the current regime i.e. Chapter 17 of Agenda 21 which identifies ocean governance—sustainable development and integrated management—as one of the essential components of the global life-support system. However, the reality is that the initiatives to manage human activities in this area show weaknesses: jurisdictional gaps and overlaps, lack of inter-agency communication and coordination, competition for scarce management resources, and inter-agency and intergovernmental conflicts. The lessons learnt from the Canadian and U.K. experiences demonstrate that ocean policy may arise either through legislative action, or an executive initiative. In each case, it is necessary that a policy exists at the highest political level all the while allowing for the participation of not only public entities but also

a stakeholder approach in business is by no means an easy and simple step to make, and rather constitutes a daily challenge for managers.[34] This in the light of the shipping industry might be more perplexing because of jurisdictional problems. A sensitive question in terms of the Arctic is—who are the main stakeholders in areas of the NWP? Then again, should the shipping industry be accountable to all non-state stakeholders of the Arctic?

It is clear that the stakeholders of the Arctic are likely to develop a different understanding of what CSR means, what they expect from the shipping industry in relation to CSR and CSP. But to burden the shipping industry with redundant liabilities is not the way to strike the much needed balance in the Arctic. Then again, the geographical environment in which this extended CSR accountability operates has changed and is changing rapidly. In an endeavour to understand the real stakeholders of the Arctic, would it be easier to categorise them as has been done in business studies for a considerable period of time. The scholars of business studies, to avoid giving the term "stakeholder" a too broadly inclusive scope, have proposed sub-categories. The most widely used is the "external/internal stakeholder framework".[35] Then again, others prefer the "primary/secondary stakeholder framework"[36] while others favour the "social/non-social stakeholder framework".[37] In addition, the commonly identified stakeholder groups include employees, local community, interest groups and civil society representatives. Some of these terms, according to scholars, by themselves raise significant problems in relation to the value of organizational accountability to stakeholders, especially "society-at-large" and the notion of community.[38] For the Arctic there exists

Footnote 33 (continued)
the private sector, scientists, academics, and others. Although for fact we are faced with a vacuum of literature which connects the regulatory regime implemented by International Organizations with that of CSR-Maritime, several initiatives have been taken that addresses more holistically the environmental aspects in the maritime industry, such as the Sustainable Shipping Initiative, the Blue Angel Logo, the Clean Shipping Index or the ISO 14001, etc. Most authors are convinced that the priority should be to create a regulatory framework that has the mandate, human and financial resources and the political will to put into practice the concept of integrated management for sustainable ocean governance. Hence, it is clearly deduced that the issues of ocean governance need to be resolved by concerted action of all stakeholders with an interest in maritime activities.

[34] Wood, D.J., "Corporate social performance revisited", *The Academy of Management Review*, 1991, Vol. 16, No. 4, pp. 691–718.

[35] Johnson, G. and Scholes, K., *Exploring Corporate Strategy: Text and Cases*, 2002, 6th edition, Harlow: FT/Prentice Hall.

[36] Weiss, J.W., *Business Ethics: A Stakeholder and Issues Management Approach*, Mason, 2003, OH: South-Western, Thomson Learning.

[37] Post, J.E., Preston, L.E. and Sachs, S., "Managing the extended enterprise: the new stakeholder view", *California Management Review*, 2002, Vol. 45, No. 1, pp. 6–28.

[38] *Supra* note 34.

several interest groups including the local indigenous inhabitants. This is the constant factor. To make things more complicated, the environment is sometimes quoted as a stakeholder. This basically engenders problems in identifying a spokesperson with which to discuss the stakeholder's concern, interests and demands.[39] Based on these theories, it is unlikely that there will be a stakeholder conglomeration for the Arctic anytime soon. This renders the "stakeholder theory" and the determination of stakeholders in terms of the Arctic as inconclusive.

---

[39] *Ibid.*

Chapter 7
Analysis of Alternatives to Set the "Balance"

Whatever is reiterated from the technological aspect, the reality of it all is that climate change is a phenomenon that will not shift gear. The Arctic, in fact, will never revert back to its original state. However, the detrimental effects of it can be lessened and it is inevitable that the present legal regime needs to change to adapt to the new situation. An analysis is necessary to extract and understand the suitable alternative that adapts to the current Arctic regime that balances different interest groups.

7.1 "Hard Law" Versus "Soft Law"

Does the vacuum of treaty law constitute the detriments of an important possibility to adequately deal with the environmental issues existing in the Arctic, or is the *status quo* soft law approach the best that could be achieved in terms of the varied interests and competence of the Arctic states?[1] The *status quo* "negotiation-dependent" framework in the Arctic, as embodied by the Arctic Council, is characterized by a "soft law" approach, which in its entirety is essentially a voluntary approach. In other words, it portrays the lack of a more strenuous "treaty approach", i.e. "hard law".[2]

---

[1] *Supra* note 9 in Chap. 4.

[2] WWF, *A New Sea: The need for a regional agreement on management and conservation of the arctic marine environment*, published by WWF International Arctic Programme, Oslo, 2008, available at; www.vliz.be/imisdocs/publications/130977.pdf (date accessed 13 April 2012).

T. Johansson and P. Donner, *The Shipping Industry, Ocean Governance and Environmental Law in the Paradigm Shift*, SpringerBriefs in Law,
DOI 10.1007/978-3-319-12541-1_7

7.1.1 The "Hard Law" Approach

The theory of incorporating a mandatory legal regime binding on the Arctic States built on the concept of the Antarctic-treaty system has been voiced by various civil-society Organizations namely, the IUCN and the WWF.[3] The WWF places forth that the Arctic needs a concrete management framework, and that such a framework must be comprehensive and ecosystem-based in order to be operative.[4] A multilateral regional treaty enshrined with provisions of management and protection of the Arctic would certainly provide a significant outline for a holistic management of the Arctic Ocean and of the ever-expanding maritime actions that are likely to shape the future of the Arctic geographical region. An all-embracing treaty would, moreover, evoke substantial diplomatic-commitments from the Arctic governments as regards sustainable development objectives coupled with serious obligations to protect the environment via pragmatic enforceable aims and agendas.[5] Many of the building blocks for implementing such "hard law" is already in place whereby, UNCLOS provides for an international framework for establishing "protected area(s)" while the Arctic Council fills in the gaps with political framework. UNCLOS in this regard is armoured with strong mechanisms in order to preserve its universality, whilst simultaneously being replete with references to rules and co-operation on the regional level.[6] The Arctic States may adopt a legally binding "Arctic Operation Scheme" which gives reference to implementing an umbrella convention i.e. UNCLOS where the Coastal States could depend on Article 234 to implement stringent control from vessel-pollution in the

---

[3] Olav Schram Stokke, "A legal regime for the Arctic? Interplay with the Law of the Sea Convention", *Marine Policy* Vol. 31, pp. 402–408, *See* also *Supra* note 1 in Foreword, where the author has focused on the Antarctic treaty which might hold the key to the solution for the problem subsisting in the Arctic. The author adds that "[a]lthough it is tempting to look to the past for solutions to the Arctic conundrum, no perfect analogy exists. The 1959 Antarctic Treaty, which froze all territorial claims and set aside the continent for scientific research, provides some lessons, but it concerns a continent rather than an ocean. Moreover, Antarctica is far removed from major trade routes, and negotiations unfolded in the entirely different context of the Cold War. As a body of water that links several large economies, the Mediterranean Sea is somewhat similar to the Arctic Ocean, but its littoral states have always had clearer historical claims, and it has never been covered with ice, at least not in human history. There is simply no comparable historical example of a saltwater space with such ambiguous ownership, such a dramatically mutating seascape, and such extraordinary economic promise." *N.B.* Richard Falk, "The Antarctic Treaty System: Are There Viable Alternatives?" in Arnfinn Jorgensen-Dahl and Willy Ostreng (eds.), *The Antarctic Treaty System in World Politics*, (London, 1991), 399 in Rothwell, 457.

[4] *Supra* note 2, where it is further stated that, it must also provide for an efficient management of human activities in the Arctic in order to conserve the living resources of the region without losing focus of sustainable development or neglecting the welfare of the traditional communities.

[5] *Ibid.*

[6] Alan Boyle, "Globalism and regionalism in the protection of the marine environment" in *Protecting the Polar Marine Environment*, Davor Vidas (ed.), 2000, Cambridge University Press, pp. 21–22.

off-shore areas. It could on the other hand provide a balance by providing restrictions on the Costal States from implementing discriminatory laws where "due regard" is paid to rights of the Flag State. In this context, it becomes transparent that the umbrella convention can moderately accommodate regional approaches to marine environmental protection subject to the condition that regional arrangements will not be inconsistent with the object and purpose of UNCLOS as laid down in Articles 237 and 311, and are in compliance with the framework for marine environmental protection established by Part XII.[7] Then again, forces from regional conventions such as OSPAR, supplements a model for procreating a pioneering regulation of pollution control for a "multiple user" specific region.

The Arctic States may also forward a request to IMO to designate the Arctic as a "Special Area" under MARPOL 73/78 or to empower the Arctic Council to designate such areas as protected areas to provide additional safeguard values from excessive voluntary vessel-pollution emanating from commercial exploitation.[8] Thus, the Arctic states could utilize marine regionalism as a basis for the development of an umbrella convention dealing with the Arctic marine environment.[9] But to what extent the Arctic States can achieve the ambition of having the Arctic designated as a "Special Area" is unforeseeable. This has not been a part of the agenda of the Arctic Council or any other IOs that have implemented regulations for the Arctic. Nevertheless, this would be completely redundant because there exists a binding legal regime, i.e. UNCLOS, that applies to the Arctic and rather than focusing on new regimes, resources should be utilised in working with what already exists. For any drawbacks, the Arctic States should work towards strengthening it. Hans Corell is of the opinion that "[t]here already exists a legal regime functioning in the Arctic—the UNCLOS and the other treaties in force in the Arctic—and we should now focus on their implementation".[10]

Be that as it may, UNCLOS is not designed to take into consideration the geographical impact of climate change with estimated provisions on how to delimit new land areas that are yet to take shape. Since the Arctic Council plays a key role in the evaluation and implementation of the Arctic legal regime, it should after considering the geographical impact, implement within the "hard law" a corresponding special provision. This special provision shall either direct the Arctic

[7] *Ibid.*, p. 22.

[8] Since UNCLOS gives reference to Regulatory Conventions, i.e. MARPOL 73/78.

[9] *Supra* note 42 in Chap. 1.

[10] Hans Corell, "Reflections on the Possibilities and Limitations of a Binding Legal Regime", *Environmental Policy and Law*, 2007, Vol. 37, No. 4, p. 321, Reference can also be made to the fact that UNCLOS contains the specific provisions on enclosed and semi-enclosed seas as well as ice-covered areas, *See* also D.M. Johnston, "The Future of the Arctic Ocean: Competing Domains of International Public Policy", 17 *Ocean Yearbook* 2003, 596, at 600–602, where it is indicated that given the 'constitutional' nature of the UNCLOS, which for the most part codifies the present law of the sea, it is no wonder that scholarly suggestions have focused on how to apply this Convention to the very particular circumstances prevailing in the Arctic region, although there are also proposals which argue that special rules should apply to the Arctic waters.

States with delimitation problems to seek the assistance of International Tribunals to delimit the maritime boundary that already exists or shall incorporate a "wait and see" policy allowing considerable time for the offshore landscapes to permanently take shape.[11] However, it is strictly upon individual States to take their own initiatives and with the existing MOU's, the States may be reluctant. Then again, the "wait and see" policy is ambiguous since the period of anticipation is uncertain. Under these circumstances, the Arctic Council may consider strict monitoring of the existing MOU's and that it does not inhibit international navigation or innocent passage. This would to a larger extent resolve complexities and set the Arctic States to deal with marine environmental projects.

7.1.2 The "Soft Law" Approach

Proponents of the "soft law" approach dismantle the "hard law" theorem, applicable in the Antarctic, as unrealistic. It is opined that the "hard law" is misdirected, given that the two poles show more differences than similarities whereby, the Arctic consists of an ocean surrounded by continents, whereas the Antarctic is a continent surrounded by ocean.[12] Moreover, the Antarctic is void of human habitation, while the Arctic is inhabited by indigenous peoples and other local communities.[13] A significant distinguishing feature is that much of the Arctic lies under the sovereignty and sovereign rights of the Arctic States, while Antarctic sovereignty claims have been frozen for the time being, and there are thus no territorial sovereigns in the Antarctic.[14] Furthermore, collaborating on a comprehensive govern-

---

[11] *N.B. Supra* note 1 in Foreword, where the author explains the complexity of the Northern Exposure and the settlements in a pragmatic manner. The author believes that "[u]ntil such a solution is found, the Arctic countries are likely to unilaterally grab as much territory as possible and exert sovereign control over opening sea-lanes wherever they can. In this legal no man's land, Arctic states are pursuing their narrowly defined national interests by laying down sonar nets and arming icebreakers to guard their claims. Russia has led the charge with its flag-planting antics this past summer. Moscow has been arguing that a submarine elevation called the Lomonosov Ridge is a natural extension of the Eurasian landmass and that therefore approximately half of the Arctic Ocean is its rightful inheritance. Naturally, other Arctic states are responding. Norway submitted its claim for additional Arctic resources to the commission in 2006; Canada and Denmark are now doing their homework in order to present their own claims. Ottawa and Copenhagen are currently at odds over the possession of Hans Island, an outcropping of desolate rocks surrounded by resource-rich waters in the Nares Strait, between Canada's Ellesmere Island and Greenland. Even the United States, despite its refusal to ratify UNCLOS, has for the past few summers dispatched its sole icebreaker to the Arctic to collect evidence for a possible territorial claim in the event the Senate eventually ratifies the treaty."

[12] Timo Koivurova, "Alternatives for an Arctic Treaty-Evaluation and a New Proposal" in *Review of European Community and International Environmental Law* (RECIEL), 17(1) 2008, Special International Polar Year Issue, pp. 14–26.

[13] *Ibid.*

[14] *Ibid.*

ance system for the Arctic can be exorbitant, and the end result could also prove to be difficult to implement, resulting in a range of new yet complex international issues hindering global commerce.[15]

Indication on the reliance of a tailor made provision in an already existing "hard law" regime, that of UNCLOS, has been endorsed by international Organizations, where Article 234 vividly vibrates Regulations that ought to be incorporated in the Arctic. However, the international Organizations have failed to realize that the conditions set forth in Article 234 relating to "ice-covered areas" will not be consistent with the Arctic region eventually, because climate change in repercussion will undoubtedly thaw ice-caps marking a longer period of "ice-free zones". In the context of international conventions as a binding treaty, it is essential to observe the number of ratifications by Arctic States. Parties to "hard law" often make reservations or exclude themselves from the law itself in order to limit their vulnerability to new commitments as the U.S. has excluded itself from being a party to UNCLOS. More complex than that is the notion of "adopting stricter standards" as encouraged by MARPOL 73/78 where some Arctic States have adopted stringent standards while others have remained silent.[16] This conflict of law in the long run creates different jurisdictions distorting the international legal regime, which is frustrating for international trade. Furthermore, creation of a binding treaty will be faced with the same problems as any other treaty or conventions where the ratification procedure is apt to be cumbersome, and the relevant parties to such binding law are sporadically hesitant to adapt to adjustments even though circumstances may have changed. With 'soft law', it is more flexible to procure widespread assent to new rules within a short period of time since they are not legally binding. This adaptable approach results in greater sub-regional trans-border co-operation and could thus encompass a wide spectrum of Arctic interests outside of the purely formal governmental level.[17]

The "soft law" approach can be virtually criticized from a number of aspects[18]; nevertheless, the primary question that revolves around this topic is the matter of "competence". The Arctic States only have competence to deal with matters over which they have control. Since the environmental problems that have emanated are not primarily generated in the Arctic, the obvious question is whether the Arctic States through the norms of "soft law" are able to undertake and mitigate environmental problems that are generated globally where they have failed the initial task of compromising in maritime delimitation issues. The solution, apparently, lies in the fact that the existing circumpolar network, to some extent, is in need of global participation.[19] But an argument which can be advanced against this is that only

---

[15] *Supra* note 2 in Chap. 4, p. 8.

[16] *Supra* note 49 in Chap. 1.

[17] *Supra* note 41 in Chap. 1, pp. 254–255.

[18] *See* para 6.2.1, p. 51.

[19] *Supra* note 10.

45 members of the 191 UN member States are parties to the Antarctic treaty.[20] Whether or not the voluntary-regional arrangements can balance the interests lies with the circumpolar political will and scientific input needed to make such measures work. Drafts of mere voluntary rules cannot solve any of the problems facing the Arctic region.[21]

7.2 Recommendations to Balance the Interests

The bearing of the last pristine eco-system accompanied by great prospects of a short-route with natural resources under troubled waters provide a two-fold purpose for the Arctic legal regime, namely to balance environmental concerns with the rapid growth of International trade.[22] The *status quo* infrastructure for marine environmental conservation in the Arctic is active at multiple levels and could be well described as a cross-road of international instruments (and referred Regulatory conventions), regional co-operation and domestic legislation. Instead of an adequate synergy or interplay between the various layers, there exists insufficient synthesis and an overlap between the international instruments and regional regulations. While this overlap continues, the Arctic States have implemented national legislation to establish sovereignty which is accomplished by incorporating unreasonable standards to the detriment of the shipping industry.[23] This entire scenario is burdened with an accelerated strain on the Arctic region as a result of climate change which needs to be dealt with decisively. It is clear that the solution lies in the regional co-operation which should be at the mid-point connecting international law at one hand, guiding

[20] *Ibid.*

[21] *Supra* note 6, p. 33.

[22] Alexander Klepikov, Alexander Danilov and Victor Dmitries, "Consequences of Rapid Arctic Climate Changes" in *International Energy Policy, The Arctic and the law of the Sea*, Myron H. Nordquist, John Norton Moore and Alexander S. Skaridov (eds.), 2005, Martinus Nijhoff Publishers, p. 227.

[23] *Supra* note 21 in Chap. 1, *Note* that, the success of International Conventions depend on actions carried out at the domestic level. As such, it is only natural that the domestic legislations of the Arctic States have been singled out as an essential element for the purposes of Arctic marine environmental protection. Initiatives taken on the domestic level can have an impact on the Regional or International level, and vice versa. Canada's unilateral response in the wake of the *Manhattan* incident resulted in Article 234 of UNCLOS, which is a lucid example. The AWPPA embodies "zero discharge" policy entailing "absolute liability". Compliance with "zero discharge" policy in a route for International navigation is difficult to relate to. This indirect policy of establishing territorial sovereignty takes its toll on the shipping industry. Then again, existence of a legal obligation for the Arctic states, and that this legal obligation has arisen as a result of the amalgamated effects of principles of international environmental regulations, marine regionalism and the obligations imposed on Coastal States would enable them to cooperatively manage the marine environment by UNCLOS.

domestic legislation on the other. The Arctic Council in this regard is the most logical body to be coordinating efforts, promoting and negotiating an "Arctic treaty", as well as becoming a decision making body.[24] Having established that, it comes down to the question as to what character should the "Arctic treaty" assume? It is quite lenient to comprehend the imperfections of both "hard law" and "soft law". What is thought to be a protective measure for the sensitive marine eco-system does not precisely spell good news for the commercial beneficiaries and *vice versa.*

It can be suggested that the balance can be brought by an intercross between "hard law" and "soft law", i.e. a combination of binding and non-binding arrangements that complement the environmental and economic state of the Arctic.[25] This can be done by considering the existing "soft law" based regional Arctic Council and the binding international treaties applicable therewith (or which are in an indirect manner related to the Arctic). Since the "Arctic treaty" will be operating under the Arctic Council, it is important to organize the main body of the Council. The Arctic Council can constitute a bipartite body, accommodating two interest groups.[26] The five Arctic States can constitute the "Arctic Council" which will be the main body, the participation of which is voluntary by Arctic States, and the Flag States who wish to utilize the Arctic sea routes shall form the "Arctic Council Members", the participation of which must be mandatory.[27] The "Arctic Council Members" will be a subsidiary body of the Council which is international in character as opposed to the "Arctic Council" which is regional.[28] Understanding the fact that the Arctic suffers from designated maritime boundaries, it will be the duty of the "Arctic Council" to primarily convert the existing MOU's into binding treaties so that the States concerned can jointly focus on environmental

[24] *Supra* note 34 in Chap. 1, where it is stated that, "[t]o do this would require the Arctic Council to shift gears from being a decent clearinghouse, but a rather weak international organization with limited funding, to an organization that has much more of a presence in domestic and international programs…."

[25] *Ibid.*, It is of course true that it is less cumbersome to replicate what has been done before, but unique problems require unique solutions. The Arctic at first has to be acknowledged as a unique environment. Although the contemporary Arctic regime is randomly connected to different branches and sources of law, nothing clearly structured and coordinated. A lot of laws are, if properly analysed, applicable to the Arctic. But it is law that differentiates between "ought" and "could" and guides the States to solutions rather than complexities. Interplay without coordination leaves more complexities than solution and that is what the Arctic is facing today.

[26] Since a balance is being sought between only two groups, namely Arctic Coastal States and States interested in International navigation.

[27] The reason for making the entire legal body partly voluntary and partly mandatory is because once the Arctic States observe that the number of International participation has increased (due to their own urgency to use a shorter route through the Arctic), the number of participation will increase, since it is feared that mandatory participation will inhibit participation.

[28] This will ensure the interplay between Regional and International regimes.

responsibilities.[29] The five working groups of the existing Arctic Council, i.e. SDWG, AMAP, PAME, CAFF and EPPR and their Secretariats will work under both the bodies where the funding for operating these working groups will be borne equally.[30]

The "Arctic treaty" will act as a coordinated and integrated nexus between the Coastal States and the Flag States with provisions which ensures the balance between both the interest groups. The Arctic Coastal States who have already ratified the international instruments must act according to the provisions (of prescriptive and enforcement jurisdiction) embedded therein. As for the States who are not parties to those international conventions will be bound by "separate provisions" which will be similar to the international regime and give "due regard" to the rights of the Flag States in terms of innocent passage and international navigation. To avoid complexity in creating two separate regimes for parties and non-parties to the international conventions, there can be one single regime parallel to the international conventions which suits the contemporary Arctic environment. This regime should accommodate "alteration provisions" that focus on climate change and the best possible measures that should or can be taken by the Coastal States in the near future to fit the situation that exists then.[31] This regime should act as a "model regime" for Arctic States to implement in domestic legislation or to amend the *status quo* national legislation accordingly. This "model regime" will help lessen "conflict of laws" and assist the Arctic States in developing a form of national alignment.[32] It should under no circumstances encourage a "zero-discharge" policy modeled after the Antarctic treaty, because in the Antarctica, there was never any need to strike a balance between environmental protection and

[29] In an effort to trace back to the geographical problems as analysed in Chap. 3, it is significant to address certain boundary issues before the Arctic States commence environmental coordination. In this proposal to create two legal organs under a bi-partite system, the main body should have a strong sense of connection so as to deal with other major problems collectively. In a way, if the existing MOU's have been ignored and violated by opposing States, there is a probability that the same may occur at any time in the future. The binding treaties will safeguard the diplomatic connection between Arctic States so that it is not weakened by any sense of violation. It is suggested that the Arctic States resolve one problem at a time and a disorganized council with an array of problems will not be the proper body to maneuver environmental strategies. There will always remain a silent problem if the maritime boundaries are not delimited or undecided under a binding treaty. If so, the Arctic council will not be able to act as the strong platform it so desires.

[30] Since funding and budgeting for these working groups have been raised as a complex issue, the financial co-operation of the non-Arctic States shall lift the burden from the Arctic States. The Coastal States may impose a moderate tariff on the Flag States, a part or its entirety (depending on discussions between both the participants) shall be utilized to operate the five working groups.

[31] i.e. Based on scientific evidence.

[32] It cannot be expected that the Arctic States shall develop similar national legislation, but when the "Arctic treaty" discourages certain policies that hinder international trade, it will act as a bar for those States in implementing such provisions of "absolute liability".

commercial interest.[33] Rather, there can be a legal framework which refers to a "multiple user" MPA so that it can set a standard of emissions or discharge of toxic elements.[34] This standard must be negotiated with the Flag States so that the standard is not unreasonable, and the balance can be maintained.[35] If, however, certain zones are considered extremely sensitive, then the Arctic States may designate them as "strict MPA(s)" restricting any amount or fragment of pollutant. However, in order to maintain equilibrium with the Flag States' smooth navigation, the Coastal State in its immediate vicinity must establish port reception facilities to assist them. There should be separate Guidelines and Regulations under the "Arctic treaty" which embody provisions as regards Flag State prescriptive and enforcement jurisdiction where flag States are under an obligation to take all precautionary measures to abide by the standards set forth in the regime itself.[36] This will be effective for both voluntary vessel-pollution and intentional vessel-pollution. Since the five Arctic countries are parties to the LC72 Convention on ocean

---

[33] Rather than replicating the Antarctic treaty regime, the Arctic has a major potential of becoming a unique model regime.

[34] This will include areas within and beyond national jurisdiction with the permission of individual Arctic States. The Arctic requires a 'multiple user' MPA in coastal and offshore environments resulting in that a single MPA can comprise a mosaic of management coupled with restriction categories in terms of voluntary marine-pollution. The primary responsibility for the designation of MPAs falls to individual States enacting specific legislations, but the Arctic area also requires MPAs beyond national jurisdiction which will be the focal point of commercial exploitation. Although MPAs provide a wide range of application, it must be acknowledged that they do not provide sufficient protection against all threats resulting from voluntary vessel-pollution. *See Supra* note 42 in Chap. 3, where Aldo Chircop has addressed many of the current MPAs as "desktop MPAs" and hence, in considering a strict protected area in the Arctic would not be of pragmatic value. According to Simon Jennings, the term "protected areas" is currently being used as "areas where there have been measurable changes in human pressure" rather than "a geographically defined area which is designated or regulated and managed to achieve specific conservation objectives." "Human pressure" in this instance can be referred to the routes through which future commercial navigation shall take place at a very fast pace creating pressure on the marine environment through intensive dumping and operational discharges. Taking this into account and to designate "any marine area" in the coastal or offshore environments and areas beyond national jurisdiction as MPA there needs to be a development of a circumpolar network in the Arctic which is complicated because of the mix of national and international agencies and jurisdictions with responsibilities in the region. In the Antarctic, the Antarctic Treaty Consultative Parties have estimated the areas where measurable changes have been caused due to human pressure and designated MPAs around sub-Antarctic islands within the greater Antarctic region. The Arctic States have to endeavor to establish a jointly agreed and coordinated network of protected areas where consistent if not identical rules for conservation and access will have to be formally implemented. Currently there are no representative channels of Marine Protected Areas in most or all of the Arctic marine area.

[35] The standards so fixed can be communicated to the IMO for approval and a request to implement them in MARPOL 73/78 for international acceptance. This is also an inherent reason for aggregating the non-Arctic States as a part of "Arctic Council". This will add to the force of the requests pledged by the Arctic States.

[36] This can be parallel to IMO Polar Shipping Guidelines.

dumping,[37] ratifying this Convention can be made mandatory for the "Arctic Council Members". Moreover, since the Arctic States are under joint contingency planning arrangements under OPRC as regards accidental oil pollution,[38] the "Arctic Council Members" will also need to be parties to the convention or form a common contingency plan with the "Arctic Council". If the "Arctic Council" is of the opinion that the international conventions relating to pollution-liability is insufficient then, the liability facet should also be incorporated in a separate "liability regime" where both the "Arctic Council" and the "Arctic Council Members" must negotiate on compensation issues for areas beyond national jurisdiction of Arctic States.[39]

[37] *Supra* note 59 in Chap. 1.

[38] *Supra* note 69 in Chap. 1.

[39] 'This liability facet' shall not include negotiations on compensations already assigned in domestic legislation of Arctic States for areas within national jurisdiction. Individual Arctic States shall maintain *status quo* of the liability regime already in force for those areas. Areas within national legislation should only be modified to the extent where it inhibits the smooth operation of international navigation with extreme standards.

Chapter 8
Conclusion

The proposed[1] Arctic legal regime, firstly, is a conglomeration of two different entities. The former entity is the congregation of the five Arctic States and the latter is the international group that is an addition to the already existing Arctic Council. Broadly interpreted, it cannot be the single concern of the Arctic states to protect an environment that has the possibility of being contaminated by an act that is global in nature. Reverting back to the analysis of the "functional criterion"[2] of the NWP, it was clear that there were "69 non-Canadian transits"[3] and the conclusion drawn from the "potential use" of the NWP was that climate change would certainly overturn this figure in the nearest decade.[4] The international community rejects the so-called "Sector theory" applied by both Canada and the Russian Federation in order to establish a right in the Arctic sea routes for prospects of commercial benefit. But it ignores the fact that, the benefits procured from the international navigation in this route may have consequences against sustainable development of that region and the Arctic Council's antiquated policy is not sufficient to combat the detrimental effects of marine contamination, which constitutes a part of the reason for such sovereign claims. Hence, the intention to bind the international community in a body[5] separate from the Arctic Council in an organizational framework, where provisions to safeguard the marine environment are weighed opposite to commercial interest in a perfect balance. This balance in the organizational structure simultaneously accommodates the interplay of international law and regional law coupled with directives guiding the national law. But the question remains as to whether the Arctic Council will be able to

[1] As proposed in para 7.2.

[2] *See* para 3.1.3, where a decisive factor in the decision of the *Corfu Channel Case* was to determine if the Strait was used for international navigation.

[3] *Supra* note 35 in Chap. 3.

[4] *Supra* note 37 in Chap. 3.

[5] i.e. "Arctic Council Members".

T. Johansson and P. Donner, *The Shipping Industry, Ocean Governance and Environmental Law in the Paradigm Shift*, SpringerBriefs in Law,
DOI 10.1007/978-3-319-12541-1_8

preserve its regional status or will this proposed system constitute another distortion in the legal regime. Since the body of "Arctic Council Members" is a subsidiary body of the "Arctic Council", it will act as an added force to the original regional cooperation based on "soft law". The logic behind the system of incorporating "Arctic Council Members" as a subsidiary body is to ensure that the international community will not have any control over the "Arctic Council" decision making authority. The "Arctic treaty" or the "single regime parallel to the international conventions" guarantees that the "Arctic Council" will not inhibit the smooth operation of vessels representing Flag States,[6] and reliance in this regard can be made on the current international instruments on Flag States' prescriptive and enforcement jurisdiction.[7] The "Arctic Council Members" will have the authority to call for negotiations when certain provisions exceed the *status quo* international *standards*. This "negotiation" is a form of silent "veto" which indicates "non-compliance" towards extreme standards thus, effective against any distortion of the existing international regime.

The new Arctic legal regime, secondly, constitutes a concrete form of stability among the Arctic States by converting the several MOUs that exist today into a single binding treaty or several binding treaties by the application of the 1969 Vienna Convention on the Law of Treaties.[8] The basic principles of *pacta sunt servanda* will be applicable whereby, the zones of conflict[9] will be governed by the rules of the treaty and the concerned States will be obliged to do so in good faith. Moreover, the new Arctic legal regime could freeze all the conflicting claims and designate those areas as a "common heritage" for the Arctic States and other stakeholders. The underlying reason is that, the geographical claims have a subtle connection with the sustainable development of the marine environment.[10] As stated in the report of the WCED as regards sustainable development, "[a] process of change in which the exploitation of resources, the direction of investments, the orientation of technological development, and institutional change are all in harmony and enhance both current and future potential to meet human needs and aspirations."[11] This binding treaty is an agreement followed by a liability facet, which will bring cohesiveness among Arctic States in the absence of any initiatives to resolve the maritime delimitation by an international Tribunal.

---

[6] In areas beyond national jurisdiction.

[7] *Supra* note 12 in Chap. 1.

[8] Malcolm N. Shaw, *International Law,* 2008, 6th Edition, Cambridge University Press, pp. 94–95, *Note* that, It depends on the intention of the five Arctic States whether the binding treaty should be bilateral or multilateral.

[9] As a result of overlapping claims.

[10] Once the claims have been resolved or joint development plans are undertaken for unresolved areas, the Arctic States can concentrate on outlining the "Arctic treaty" provisions to control international shipping for the Arctic sea routes combined.

[11] World Commission of Environment and Development, *Our Common Future*, 1987, New York: Oxford University Press, pp. 43–46.

The new "Arctic Council" is actually functioning as that "institutional change", which is a part of the big "process of alteration" in the sustainable development mechanism. The question revolves around the issue as to what extent the Arctic States are prepared to harmonise issues among them to bring the final balance between the "Arctic Council" and "Arctic Council Members". Although the Russian Federation's placement of a flag at the North Pole acted as an undermining statement on the face of mutual agreement, the two existing MOUs in the NSR are certainly praiseworthy and reflects the intention to bring the anticipated "harmony".[12] To date eight bilateral maritime boundaries have been delimited and there are several more that will need to be delimited via regional treaty.[13] The Arctic States should be united and prepared for arrangements to strengthen the "Arctic Council" both in terms of geographical and political aspects, which will in the long run have an influence over non-Arctic States navigating through the Arctic sea routes. This binding treaty, moreover, will provide a firm framework for co-operation among Arctic States under OPRC and strengthen the established joint contingency arrangements to combat oil pollution.[14] With the binding treaty as a root for co-operation, the "Arctic Council" can adopt mandates to transform the environmental segment of the Polar Shipping Guidelines into a set of Directives to impose an *obligation*[15] on the Flag States in respect of operational vessel-source pollution. Hence, the non-legal binding nature of the Polar Shipping Guidelines can be given a binding reality in the new legal regime governed by the "Arctic Council".

On the opposite side of the equation stands the interest of the shipping industry. What started with an inter-Arctic navigation of the *Manhattan* voyage, has evolved as new routes connecting trading partners on the premises of the warming and ice melting Arctic ocean, which is envisaged as being faced with accelerated ship traffic. The NWP and NSR have emerged out of the *three*[16] main routes because of the support that the Coastal States can provide for international shipping.[17] Nevertheless, the international shipping in turn needs to be within the conglomeration of the Council since it is the only body that seeks the so-called "commercial interest". "Observer" and "*Ad hoc* observer" status is already open for non-Arctic States.[18] The analysis is that,

[12] *Supra* note 81 in Chap. 3.

[13] *Supra* note 34 in Chap. 1, p. 205.

[14] *Supra* notes 69 and 70 in Chap. 1.

[15] In the form of discharge restrictions or indicate "Sox emission control areas".

[16] The three main routes include Arctic Ocean Proper, the NSR and the NWP.

[17] *Supra* note 34 in Chap. 1, p. 206.

[18] *N.B.*, Observer status is open to non-Arctic states approved by the Council at the Ministerial Meetings that occur once every two years. Permanent observers have no voting rights in the Council. As of May 2013, twelve non-Arctic states have Permanent Observer status. Then again, Ad hoc observer states need to request permission for their presence at each individual meeting; such requests are routine and most of them are granted. There are six ad hoc members, not including the European Union. At the 2013 Ministerial Meeting in Kiruna, Sweden, the EU requested full observer status. It was not granted, mostly because the members do not agree on the EU ban on hunting seals.

if the Coastal States give "due regard" to international navigation as endorsed by Article 234 of UNCLOS,[19] the equation is correctly balanced if the international sector has the same "due regard" for the marine environment of the Arctic Coastal States. But the term "due regard" may have different interpretations in terms of different layers of the maritime zones and may be debated in terms of degree and level.

Regulations and Guidelines in the form of "hard law" can be an option to ensure that the Arctic marine environment is within the Flag State agenda. Questions can be raised as to why the participation of individual Flag States is made compulsory with compelling Regulations. While "soft law" may be helpful, their application pertaining to Flag States will never be uniform because of its non-binding legal nature. International "hard law" instrument or a regional "hard law" instrument that is international in character would set common standards of enforcement and would work towards eliminating the difficulties in events where each Flag State applies different standards when issues relating to operational discharge, voluntary discharge or deliberate dumping emanate. But international "hard law" instrument is a maritime myth and has never existed. However, the keyword of "hard law" is definite accountability and to ensure this in the field of the marine environment, there needs to be enforcement of binding Regulations. Individual accountability can be ensured if the participation of individual Flag States is made mandatory and is adjoined in the taskforce of sustainable development policy making. If the Arctic States should deter from a "zero discharge" policy corresponding to absolute liability, they must be ensured with agreements that the pristine environment will be free from hazards of Arctic navigation. The Arctic Coastal States, however, for enforcement of "hard law" must satisfy certain conditions to ensure efficacious compliance. The construction of port facilities in the Arctic has been extremely limited and is a main constraint for the provisions of OPRC to operate effectively.[20] To help Flag States comply with "hard law", the "Arctic Council" must establish adequate ports, monitor Flag State navigation via Port State control and provide port reception facilities.[21] But then again, extreme regulations have always been subject to criticism at large. Directive 2005/35/EC of the EU states that ship-source polluting discharges constitute in principle a criminal offence and according to the Directive this relates to discharges of oil or other noxious substances from vessels.[22] The main principle which guides this Directive is that the existing Civil Liability Convention is not enough to deter pollution in the waters of the EU. As such, drastic measures were necessary and this "hard law" system has invoked other legal issues i.e. violation of human rights. But on the positive side, this has made the shipping industry more aware of discharge of

---

19 *Supra* note 25 in Chap. 1.

20 *Supra* note 69 in Chap. 1, p. 155, where it is indicated that emergency response is particularly challenging in the Arctic for a variety of reasons, including the remoteness and great distances that are often involved in responding.

21 *Ibid.*

22 Directive 2005/35/EC of the European Parliament and of the Council of 7 September 2005 on ship-source pollution and on the introduction of penalties for infringements.

pollutants in the EU waters and inspite of extreme objection from the shipping industry—the Directive is constant in its wordings. Could this be a replica of the innovative "hard law" for the Arctic or could it be just another legal mistake?

The bipartite concept, like every organization can be subject to criticism and therefore, disputed. However, when disputed it must be recalled that the issue to be dealt with is climate change in the Arctic and the distortion it is causing to the existing legal regime. It is not only a matter of balancing certain interests, but it is adjusting "soft law" and "hard law" where flexible policies can meet accountability. In between, the moderate tariff imposed on the Flag State will be sufficient to fund the five subsisting working groups, and the secretariat of the five working groups will collaborate with the Flag States in trying to develop strict MPA's.[23] Not only does this solve the budget predicament, but also allows transparency insofar as the Flag State has an equal voice in the decision, which does not override the Council's decision. In this hypothetical model, the national legislation of Canada[24] and the U.S.[25] could serve as an inspiration, and could play a pivotal role in development of an "Arctic treaty" via the Arctic Council. A joint proposal on behalf of the "Arctic Council" can be forwarded to IMO, stating that areas designated by the Council as MPA along with discharge standards as negotiated between the Arctic States and international participants be referred in MARPOL 73/78. This will save the Council from the cumbersome process of creating a separate legal regime and the Arctic States can insist ratification of MARPOL 73/78 on non-Arctic members whereby the other detailed Annexes may assist in future references for environmental policies which is inevitable as a result of climate change. Then again, when a certain area is ascertained as an MPA, the limits of pollution discharge for those areas automatically declines.[26]

---

[23] *Supra* note 30 in Chap. 7.

[24] *Supra* note 22 in Chap. 2.

[25] *Supra* notes 68 and 70 in Chap. 2.

[26] *N.B.* Official website of Fisheries and Oceans Canada, available at; http://www.dfo-mpo.gc.ca/oceans/marineareas-zonesmarines/mpa-zpm/index-eng.htm (date accessed 6 August 2014), where it is stated that "[e]stablishing MPAs within the context of integrated oceans management provides a mechanism for taking into account stakeholder input as well as broader ecological, social, cultural and economic considerations. It also provides an opportunity to reinforce conservation measures with complementary management regimes implemented in surrounding areas, including linkages with broader ecosystem objectives, as well as land-based initiatives such as habitat protection and enhancement, pollution control, land use controls and the establishment of coastal terrestrial parks. This approach of nesting MPAs within broader planning initiatives helps maintain the integrity and long-term viability of the MPA and maximize the conservation effectiveness of all MPA planning processes." This is also aligned with the concept of PSSA undertaken by IMO whereby PSSA has been defined as "… an area that needs special protection through action by IMO because of its significance for recognized ecological or socio-economic or scientific reasons and which may be vulnerable to damage by international maritime activities. The criteria for the identification of particularly sensitive sea areas and the criteria for the designation of special areas are not mutually exclusive. In many cases a Particularly Sensitive Sea Area may be identified within a Special Area and vice versa".

Sustainable development and global partnership are intertwined. It is after all, the arguments in favour of international navigation that has led to the concept of balancing interests. If the proposed legal regime is put aside and "refinement" or "revision" of the existing regional approach is considered, then the conflict of laws will still be constant. The Arctic States in an effort to uphold sustainable development behind the facade of state sovereignty will not deter from extreme standards. Objections will be aimed at Arctic policies by the shipping industry and the Arctic legal regime will continue to vibrate the absence of global partnership. The preamble of Agenda 21 addresses the need for global partnership and highlights the significance of a balance and integrated approach, "…[i]t reflects a global consensus and political commitment at the highest level on development and environment cooperation. Its successful implementation is first and foremost the responsibility of Governments International cooperation should support and supplement such national efforts. In this context, the United Nations system has a key role to play. Other international, regional and subregional organizations are also called upon to contribute to this effort. The broadest public participation and the active involvement of the non-governmental organizations and other groups should also be encouraged."[27] This, of course, should be combined with decisive action based on structure and coordination. The Arctic States must with the valid participation of international community set an example and portray to the world at large that it is possible to create, agree upon and implement an efficient management regime that takes both environmental protection international trades into account. Other regimes are not worth following when the Arctic can be established as a unique model itself. Then again, environmental protection is a sensitive issue for the Arctic States and only upon participation in a common legal body can the international community comprehend this necessity.

No matter what configuration the new Arctic legal regime ends up taking in the end, it must accommodate both international navigation and the concept of sustainable development. International navigation must come with accountability and sustainable development must be accompanied by "due regard" for international trade and commerce. This balance will always remain in theory unless the

[27] *United Nations Conference on Environment & Development*, Rio de Janeiro, Brazil, 3–14 June 1992, Agenda 21, United Nations Sustainable Development, *See* Chapter 1, Preamble para 1.3, *See* also Chapter 2, para 2.1, where it is stated that, "…[t]his partnership commits all States to engage in a continuous and constructive dialogue, inspired by the need to achieve a more efficient and equitable world economy, keeping in view the increasing interdependence of the community of nations and that sustainable development should become a priority item on the agenda of the international community. It is recognized that, for the success of this new partnership, it is important to overcome confrontation and to foster a climate of genuine cooperation and solidarity. It is equally important to strengthen national and international policies and multinational cooperation to adapt to the new realities." This if related to the Arctic deduces the fact that partnership between "Arctic Council" and "Arctic Council Members" can surely foster the desired cooperation to balance interests. Global partnership has, thus, been acknowledged as a catalyst that can bring a balance in the uneven equation.

concerned groups connected to each of these categories are in a common platform.[28] The global participation must be followed by a legal order, an order which is necessary to combat the effects of climate change at the turn of the century incorporated under a defined body. They should "break the ice" and act fast, before climate change breaks the polar ice thus, leading to confusion, complexity and disarray.

[28] *Ibid.*, *See* Chapter 2, para 2.8, where it is stated that, "[t]he international trading environment has been affected by a number of developments that have created new challenges and opportunities and have made multilateral economic cooperation of even greater importance."

Table of Cases

30 June 1865	*Aves Island* Case, Arbitral award relating to the issue of control and sovereignty over Aves island, raised between Venezuela and the Kingdom of the Netherlands, Vol. XXVIII, pp. 115–124
21 April 1870	*Island of Bulama Case*, Arbitral Award between Portugal and the United Kingdom, Reports of International Arbitral Awards, Vol. XXVIII, pp. 131–140
6 June 1904	*Guiana Boundary Case*, Brazil versus Great Britain, Reports of International Arbitral Awards, Vol. XI pp. 11–23
4 April 1928	*Island of Palmas Case*, United States of America versus The Netherlands, Permanent Court of Arbitration Rep., The Hague, p. 837
5 September 1933	*Legal Status of Eastern Greenland*, Denmark versus Norway, Permanent Court of International Justice, (ser. A/B) No. 53 (Apr. 5) Judgment No. 20, Publications of the Permanent Court of International Justice Series A./B. No. 53; Collection of Judgments, Orders and Advisory Opinions, A.W. Sijthoff's Publishing Company, Leyden, p. 732
9 April 1949	*Corfu Channel* Case, The Government of the United Kingdom of Great Britain and Northern Ireland versus the Government of the People's Republic of Albania, (Merits), Judgment of the International Court of Justice, ICJ Rep. p. 4
18 December 1951	*Anglo-Norwegian Fisheries Case*, United Kingdom versus Norway, 18 December, International Court of Justice Report, p. 116
17 November 1953	*The Minquires and Ecrehos Case*, International Court of Justice Reports, p. 47, 52, 67
16 October 1975	*Western Sahara Case*, Advisory Opinion, International Court of Justice Reports, p. 12
27 June 1986	*Nicaragua Case,* Nicaragua versus US, International Court of Justice Reports, pp. 14–101, para. 213

T. Johansson and P. Donner, *The Shipping Industry, Ocean Governance and Environmental Law in the Paradigm Shift*, SpringerBriefs in Law,
DOI 10.1007/978-3-319-12541-1

Bibliography

Books

Anderson, Alun, *After the Ice: Life, Death, and Geopolitics in the New Arctic*, 2009, New York.
Berger, W.H., Smetacek, V.S. and Wefer, G., *Productivity of the Ocean: Present and Past*, 1989, John Wiley and Sons, New York.
Birnie, Patricia and Boyle, Alan, *International Law and the Environment*, 2002, Oxford University Press, 1995, Cambridge, M.A.
Chayes, Abram and Antonia Handler Chayes, *The New Sovereignty: Compliance with International Regulatory Agreements.*
Churchill, R.R., and Lowe, A.V., *The Law of the Sea*, 3rd Edition, 1999, Manchester University Press.
Davis, K. and Blomstrom, R. L., *Business and its Environment*, New York: McGraw Hill.
Doerffer, J., *Oil spill response in the marine environment* (1st ed.), 1992, Oxford: Author.
Doelle, Meinhard, *From Hot Air to Action? Climate Change, Compliance and the Future of International Environmental Law*, 2005, Thomson/Carswell.
Elverland, Ellen, *The Arctic System: International Polar Year 2007/2008*, Norwegian Polar institute, 2009, ISBN 827666251X, 9788276662511.
Frank, Veronica, *The European Community and the Marine Environmental Protection in the International Law of the Sea: Implementing Global Obligations in the Regional Level*, 2007, Martinus Nijhoff Publishers.
Gold, Edgar, *Handbook on Marine Pollution* (2nd ed.), Assuranceforeningen Gard, 1998, ISBN 82-90344-11-2, Arendal.
Gold, Edgar, *Gard Handook on Protection of the Marine Environment*, 3rd Edition, 2006, Arendal: Assuranceforeningen Gard.
Gunderson, L.H. and C.S. Holling (eds.), *Panarchy: Understanding transformations in human and natural systems*, Washington, 2002, DC.
Heap, John (Editor), *Handbook of the Antarctic Treaty System*, 1994, 8th Edition.
Hunter, David, Salzman, James and Zaelke, Durwood, *International Environmental Law and Policy*, Foundation Press: New York, 1091.
Huybrechts, Marc A., *Criminal Liability of Master and Crew in Oil Pollution Cases*, Maritime Pollution Liability and Policy: China, Europe and the US, Edited by Michael G. Faure, Han Lixin & Shan Hongjun, 2010, Kluwer Law International.
Jakobsson, Linda, *China Prepares for an Ice-free Arctic*, 2010, SIPRI Insights on peace and Security, No. 2010/2.
Johnson, G. and Scholes, K., *Exploring Corporate Strategy: Text and Cases*, 2002, 6th edition, Harlow: FT/Prentice Hall.

T. Johansson and P. Donner, *The Shipping Industry, Ocean Governance and Environmental Law in the Paradigm Shift*, SpringerBriefs in Law,
DOI 10.1007/978-3-319-12541-1

Jones, T.M., Wicks, A.C. and Freeman, R.E., *Stakeholder theory: the state of the art*, in N.E. Bowie (Ed.). 2002, The Blackwell Guide to Business Ethics, Oxford: Blackwell.

Kingdon, J.W. *Agendas, Alternatives, and Public Policies*, 2nd ed., 1995, New York.

Klepikov, Alexander, Danilov, Alexander and Dmitries, Victor, "Consequences of Rapid Arctic Climate Changes" in *International Energy Policy, The Arctic and the law of the Sea*, Myron H. Nordquist, John Norton Moore and Alexander S. Skaridov (eds.), 2005, Martinus Nijhoff Publishers.

Kovalev, Aleksandr A. and Butler, William, *Contemporary Issues of the Law of the Sea: Modern Russian Approaches*, 2004, Eleven International Publishing.

Krupnik, Igor, *Arctic Adaptations: Native Whalers and Reindeer Herders of Northern Eurasia*, 1993, Hanover, NH.

Litfin, Karen T., *Ozone Discourses: Science and Politics in Global Environmental Cooperation*, 1994, New York.

Mansell, John N.K., *Flag State Responsibility: Historical Development and Contemporary Issues*, 2009, Springer.

Martin Stopford, *Maritime Economics*, 2009, London: Routledge, Taylor and Francis Group.

Morrison, Anthony P., *Places of Refuge for Ships in Distress: Problems and Methods of Resolution*, Series: Legal Aspects of Sustainable Development. Martinus Nijhoff, 2012, DOI: 10.1163/9789004218888.

Nordquist, Myron H., *United Nations Convention on the Law of the Sea, 1982: A Comentary*, University of Virginia, Center for Oceans Law and Policy, Vol. 3.

O' Connell, Daniel Patrick, *The International Law of the Sea*, Vol. I, 1982, Oxford: Clarendon Press.

Osherenko, Gail and Oran R. Young, *The Age of the Arctic: Hot Conflicts and Cold Realities*, 1989, Cambridge, UK.

Ostreng, W. (Ed.), *International Northern Sea Route Programme Integration Book*, 1999, Tokyo, Japan: Ship and Ocean Foundation.

Pattberg, Philipp H., *Private Institutions and Global Governance: The New Politics of Environmental Sustainability*, Cheltenham, UK. 2007; Anderson 2009.

Peterson, M.J., *Governing the Frozen South: The Creation and Evolution of the Antarctic Treaty*, System, 1988, Berkeley, CA.

Pharand, Donat in association with Legault, Leonard H., *Northwest Passage:Arctic Straits*, 1984, Martinus Nijhoff Publishers.

Pharand, Donat, *The Law of the Sea of the Arctic: with Special Reference to Canada*, 1973, University of Ottawa Press.

Pharand, Donat, *Canada's Arctic waters in international law*, 1988, Cambridge University Press.

Querol, D., *Genetic Resources: A Practical Guide to Their Conservation*, 1983, Zed Books Ltd., London.

Roach, Ashley J. and Smith, Robert W., *United States Responses to Excessive Maritime Claims*, Martinus Nijhoff Publishers (The Hague/Boston/London), 1992, Second edition, pp. 339–353.

Rothwell, Donald R., *The Polar Regions and the Development of International law*, 1996, Cambridge University Press.

Sassen, Saskia, Territory, Authority, *Rights: From Medieval to Global Assemblages*, Princeton 2006.

Shaw, Malcolm N., *International Law*, 2008, 6th Edition, Cambridge University.

Stokke, Olav Schram, *Radioactive waste in the Barents and Kara Seas: Russian implementation of the global dumping regime* in Protecting the Polar Marine Environment, Davor Vidas (ed.), 2000, Cambridge University Press.

Tan, Alan Khee-Jin, *Vessel-Source Marine Pollution: The Law and Politics of International Regulation*, Cambridge University Press, 2006, p. 234.

Theutenberg, Bo Johnson, *The Evolution of the Law of the* Sea, 1984, Tycooly International Publishing Company.

Vidas, Davor, *Protecting the Marine Environment: Law and Policy for Pollution Prevention*, Cambridge University Press 2000.

Vukas, Budislav, *United Nations Convention on the Law of the Sea and the Polar Marine Environment*, Davor Vidas (ed.), *Protecting the Polar Marine Environment: Law and Policy for Pollution Prevention*, 2000, Cambridge University Press.

Walker, Brian and David Salt, *Resilience Thinking: Sustaining Ecosystems and People in a Changing World*, 2006, Washington, DC.

Waltz, Kenneth N., *Theory of International Politics*. 1979, Reading, MA, ISBN 0075548526.

Weiss, J. W., *Business ethics: A managerial, stakeholder approach*, 1994, Belmont, Calif: Wadsworth Pub. Co.

Weiss, J.W., *Business Ethics: A Stakeholder and Issues Management Approach*, Mason, 2003, OH: South-Western, Thomson Learning.

Young, Oran R., *Governance in World Affairs*, Ithaca, 1999, New York.

Özçayir, O., *Port State Control* (2nd ed.), LLP, 2004, ISBN 1-8431-1328-7, London.

Chapters in Books

Anderson, David, "Port States and Environmental Protection" in *International Law and Sustainable Development: Past Achievements and Future Challenges*, Alan Boyle and David Freestone (eds.), 10 May 2001, OUP Oxford.

Boyle, Alan, "Globalism and regionalism in the protection of the marine environment" in *Protecting the Polar Marine Environment*, Davor Vidas (ed.), 2000, Cambridge University Press.

Butler, William E., "The legal regime of Soviet Arctic marine areas" in *The Soviet Maritime Arctic*, Lawson W. Brigham (ed.), 1991, Belhaven Press.

Brubaker, Douglas R., "Regulation of Navigation and Vessel Source Pollution in the Northern Sea Route: Article 234 and State Practice" in *Protecting the Polar Marine Environment: Law and Policy for Pollution Prevention,* Davor Vidas (ed.) 2000, Cambridge University Press.

Chircop, Aldo, "International Arctic Shipping: Towards Strategic Scaling-Up of Marine Environmental Preotection" in *Changes in the Arctic Environment and the Law of the Sea*, Myron H. Nordquist, John Norton Moore and Tomas H. Heidar (eds.), 2010, IDC Publishers, Martinus Nijhoff Publishers and VSP.

Chircop, Aldo, "Challenges for the Regulation of International Shipping through the Arctic" in *Impacts of Climate Change on the Maritime Industry*, The proceedings of the Conference on Impacts of Climate Change on the Maritime Industry, 2–4 June 2008 (Sweden), Neil Bellefontaine and Olof Linden (eds.), 2009, World Maritime University Publications.

Daugherty, E. L., "Public Relations and social Responsibility", in R. L. Health (Ed.), *Handbook of Public Relations*, London: Sage Publications.

DeSombre, Elizabeth R. and Barkin, J. Samuel, "International Trade and Ocean Governance" in *Securing the Oceans: Essays on Ocean Governance—Global and Regional Perspectives*, Chua Thia-Eng, Gunnar Kullenberg and Danilo Bonga (eds.), January 2008, Published by GEF, UNDP and IMO in association with the Nippon Foundation.

Ehlers, Peter, "Effects of Climate Change on Maritime Transportation" in *Impacts of Climate Change on the Maritime Industry*, The proceedings of the Conference on Impacts of Climate Change on the Maritime Industry, 2–4 June 2008 (Sweden), Neil Bellefontaine and Olof Linden (eds.), 2009, World Maritime University Publications.

Elferink, Alex G. Oude, "Arctic Maritime Delimitations: The preponderance of Similarities with Other Regions" in *The Law of the Sea and Polar Maritime Delimitation and Jurisdiction*, Alex G. Oude Elferink and Donald R. Rothwell (eds.), 2000, 185 Kluwer International, Great Britain.

Emiliani, M.L., *A mathematical logic approach to the shareholder vs stakeholder debate*, Management Decision, 2001, Vol. 39, No. 8.

Falk, Richard, "The Antarctic Treaty System: Are There Viable Alternatives?" in Arnfinn Jorgensen-Dahl and Willy Ostreng (eds.), *The Antarctic Treaty System in World Politics*, (London, 1991), 399 in Rothwell, 457.

Frank, Veronica, "The European Community and Marine Environmental Protection" in the *International Law of the Sea: Implementing Global Obligations at the Regional Level*, Martinus Nijhoff, 2007, ISBN: 978 90 04 15695 1.

Frederick, W.C., *Theories of corporate social performance*, in S.P. Sethi and M.C. Falbe (Eds.) Business and Society: Dimensions of Conflict and Cooperation, 1987, Lexington, MA: Lexington Books/D.C. Heath.

Hubert, R. "Article 234 and Marine Pollution Jurisdiction in the Arctic" in Oude Elferink A.G. and Rothwell, D.R. (eds.), *The Law of the Sea and Polar Maritime Delimitation and Jurisdiction,* 2001, Martinus Nijhoff Publishers.

Hubert, R., "Cooperation or Conflict in the Arctic" in *Changes in the Arctic Environment and the Law of the Sea*, Myron H. Nordquist, John Norton Moore and Tomas H. Heidar (eds.), 2010, IDC Publishers, Martinus Nijhoff Publishers and VSP.

Joyner, Christopher, "The Legal Status and Effect of Antarctic Recommended Measures" in Dinah Shelton, ed., *Commitment and Compliance*, 2000, Oxford: Oxford University Press.

Kimball, Lee, "The Antarctic Treaty System" in John R. Twiss and Randall Reeves, eds., *Conservation and Management of Marine Mammals*, Washington: Smithsonian Institution Press, 1999.

Kennedy, Janelle, Hanson, Arthur J., and Mathias, Jack, "Ocean Governance in the Arctic: A Canadian Perspective" in *Securing the Oceans: Essays on Ocean Governance—Global and Regional Perspectives*, Chua Thia-Eng, Gunnar Kullenberg and Danilo Bonga (eds.), January 2008, Published by GEF, UNDP and IMO in association with the Nippon Foundation.

Kirton, John and Munton, Don, "The Manhattan Voyages and their Aftermath" in *Politics of the Northwest Passage*, Franklyn Griffiths (eds.), 1987, McGill-Queen's University Press.

Lalonde, Suzanne and MacDonald, Ronald St. J., "Donat Pharand: The Arctic Scholar" in *The Canadian Yearbook of International Law-Annuaire Canadien de droit international*, 2006, Vol XLIV, University of British Columbia Press.

Macko, Stephen A., "Changes in the Arctic Environment" in *Changes in the Arctic Environment and the Law of the Sea*, Myron H. Nordquist, John Norton Moore and Tomas H. Heidar (eds.), 2010, IDC Publishers, Martinus Nijhoff Publishers and VSP.

McRae, D.M., "The Negotiation of Article 234" in *Politics of the Northwest Passage*, Franklyn Griffiths (eds.), 1987, McGill-Queen's University Press.

M'Gonigle, R. Michael and Zacher, Mark W., "Canadian Foreign Policy and the Control of Marine Pollution" in *Canadian Foreign Policy and the Law of the Sea*, Barbara Johnson and Mark W. Zazher (eds.), 1977, The University of British Columbia.

Mukherjee, Proshanto K., "The Penal Law of Ship Source Marine Pollution" in *Law of the Sea, Environmental Law and Settlement of Disputes: Liber Amicorum Judge Thomas Mensah*, 2007, Martinus Nijhoff Publishers.

Ragner, Claes Lykke, "Den norra sjövägen", In Hallberg, Torsten (ed.), *Barents—ett gränsland i Norden (Norden Association's Yearbook)*, Stockholm, Arena Norden, 2008, pp. 114–127, English translation available at; <http://www.fni.no/doc&pdf/clr-norden-nsr-en.pdf> (date accessed 3 March 2012).

Raustiala, Kal and Slaughter, Anne-Marie, International Law, International Relations and Compliance. In *The Handbook of International Relations*, eds. Walter Carlnaes, 2002, Thomas Risse and Beth Simmons. London: Sage Publications, Ltd.

Reid, Robert S., "The Canadian Claim to Sovereignty Over the Waters of the Arctic", 12 *The Canadian Yearbook of International Law*, 1974.

Rothwell, Donald R. and Joyner, Christopher C., "Global environmental protection instruments and the polar marine environment" in *Protecting the Polar Marine Environment*, Davor Vidas (ed.), 2000, Cambridge University Press.

Rothwell, Donald R., and Joyner, Christopher C., "Domestic perspectives and regulations in protecting the polar marine environment: Australia, Canada and the United States" in *Protecting the Polar Marine Environment*, Davor Vidas (ed.), 2000, Cambridge University Press.

Rothwell, Donald R., & VanderZwaag, David L., "The Sea Change Towards Principled Oceans Governance" in *Towards Principled Oceans Governance: Australian and Canadian Approaches and Challenges* (London: Routledge Press, in press), 2006.

Skaridov, Alexander S., "Russian Policy on the Arctic Continental Shelf" in *Changes in the Arctic Environment and the Law of the Sea*, Myron H. Nordquist, John Norton Moore and Tomas H. Heidar (eds.), 2010, IDC Publishers, Martinus Nijhoff Publishers and VSP.

Stokke, O., "Protecting the Polar Marine Environment : Law & Policy for Pollution Prevention" in D. Vidas (Ed.), 2007, Port Chester, NY, USA: Cambridge University Press.

Solomon, A., "Management of terrestrial parks and reserves during climate change" in: Pernetta, J., Leemans, R., Elder, D. and Humphrey, S. (eds.), *Impact of Climate Change on Ecosystems and Species: Implications for Protected Areas*, 1994, International Union for Conservation of Nature and Nature Reserves, Gland, Switzerland.

Van Pay, Brian J., "National Maritime Claims in the Arctic" in in *Changes in the Arctic Environment and the Law of the Sea*, Myron H. Nordquist, John Norton Moore and Tomas H. Heidar (eds.), 2010, IDC Publishers, Martinus Nijhoff Publishers and VSP.

Young, Oran R., "Arctic Shipping: An American Perspective" in *Politics of the Northwest Passage*, Franklyn Griffiths (eds.), 1987, McGill-Queen's University Press.

International Treaty Instruments

Convention on the Territorial Sea and the Contiguous Zone, 1958, Done at Geneva on 29 April 1958, Entered into force on 10 September 1964, United Nations, *Treaty Series*, vol. 516.

DIRECTIVE 2005/35/EC of The European Parliament and of the council of 7 September 2005; on ship-source pollution and on the introduction of penalties for infringements, available at; <http://cleanseanet.emsa.europa.eu/docs/public/Directive_2005_35_EC.pdf> (date accessed 29 February 2012).

Guidelines for Ships Operating in Arctic Ice-Covered Areas, Adopted by IMO MSC/Circ. 1056, MEPC/Circ. 399, of 23 December 2002, available at <http://www.imo.org/blast/blastDataHelper.asp?data_id=29985&filename=A1024(26).pdf> (date accessed 22 February 2012).

International Convention for the Prevention of Pollution from Ships, 1973 as modified by the protocol of 1978, Adoption: 1973 (Convention), 1978 (1978 Protocol), 1997 (Protocol—Annex VI); Entry into force: 2 October 1983 (Annexes I and II).

International Convention Relating to Intervention on the High Seas in Cases if Oil Pollution Casualties (with annex official Russian and Spanish translations and Final Act of the International Legal Conference on matine pollution damage, 1969). Concluded at Brussels on 29 November 1969, available at; <https://treaties.un.org/doc/Publication/UNTS/Volume%20970/volume-970-I-14049-English.pdf> (date accessed 5 August 2014).

United Nations Convention on the Law of the Sea, Montego Bay, 10 December 1982, available at: <http://www.un.org/depts/los/convention_agreements/texts/unclos/unclos_e.pdf>.

National Legislation and Guidelines
Canada

Arctic Water Pollution Prevention Act 1970 [r.s.c. 1985 (1st Supp.) C.2, (1st Supp.) S.1.], available at; <http://www.tc.gc.ca/media/documents/acts-regulations/A-12-acts.pdf> (date accessed 29 February 2012).

Arctic Shipping Pollution Prevention Regulations, C.R.C. Ch. 353, available at; <http://laws.justice.gc.ca/PDF/Regulation/C/C.R.C.,_c._353.pdf> (date accessed 29 February 2012).

Canada Shipping Act, 2001 S.C., ch.26, available at; <http://www.tc.gc.ca/media/documents/acts-regulations/C-10.15-acts.pdf> (date accessed 29 February 2012).
Migratory Birds Convention Act, S.C., 1994, Ch. 22, available at; <http://laws.justice.gc.ca/PDF/Statute/M/M-7.01.pdf> (date accessed 29 February 2012).
Oceans Act (Consolidation), S.C. 1996, c. 31, available at; <http://laws-lois.justice.gc.ca/PDF/O-2.4.pdf> (date accessed 5 August 2014).
Territorial Sea and Fishing Zones Act, S.C. 1964-65, c. 22.

Russian Federation

Federal Law of 31 July 1998, available at; <http://www.un.org/Depts/los/LEGISLATIONANDTREATIES/PDFFILES/RUS_1998_Act_TS.pdf> (date accessed 3 March 2012).
Federal Law of 17 December 1998, available at; <http://www.un.org/Depts/los/LEGISLATIONANDTREATIES/PDFFILES/RUS_1998_Act_EZ.pdf> (date accessed 3 March 2012).
Federal Law on Environmental Protection of January 10 January 2002, N 7-FZ, available at; <http://www.icfinternational.ru/doc_files/oops.pdf> (Russian version) (date accessed 3 March 2012).
Guide to Navigating through the Northern Sea Route, Head Department of Navigation and Oceanography of the Ministry of Defense of the Russian Federation.
Maritime Doctrine of Russian Federation 2020, approved by President Vladimir Putin on 27.06.2001, available at; <http://www.oceanlaw.org/downloads/arctic/Russian_Maritime_Policy_2020.pdf> (date accessed 3 March 2012).
Water Code of the Russian Federation of 3 June 2006, N 74-FZ, available at; <http://www.icfinternational.ru/doc_files/vodn_kodeks.pdf> (Russian version) (date accessed 3 March 2012).

United States

The Act to Prevent Pollution from Ships, 33 United States Codes §§ 1901–1903, available at; <http://epw.senate.gov/atppfs.pdf> (date accessed 5 March 2012).
The Coastal Zone Management Act, 1972, as amended through Pub. L. No. 109-58, the Energy Policy Act of 2005, available at; <http://coastalmanagement.noaa.gov/about/media/CZMA_10_11_06.pdf> (date accessed 5 March 2012.
The Oil Pollution Act of 1990, 33 United States Codes § 2701–2761 (1990), available at; <http://epw.senate.gov/opa90.pdf> (date accessed 5 March 2012).

Denmark

Denmark, Greenland and Faroe Islands: Kingdom of Denmark Strategy for the arctic 2011-2020, available at; <https://web.law.columbia.edu/sites/default/files/microsites/climate-change/files/Arctic-Resources/Arctic-Council/02_01_Denmark%20strategy%20for%20the%20Arctic.pdf≥ (date accessed 29.07.2014).
Greenland Home Rule Executive order Concerning Arctic Char Fishing, This is in accordance with §23, §25, §26, §33, sect. 2 and §34, sect. 3 and §19 in Act of Landsting No. 11 of November 12th 1980 concerning Nature Preservation, available at <http://naalakkersuisut.gl/~/

media/Nanoq/Files/Attached%20Files/Engelske-tekster/Legislation/Arctic%20Char%20Fishing.doc> (date accessed 31.07.2014).

Finland

Act on Water Resources Management 1299/2004, Given in Helsinki on 30 December 2004, available at <http://www.finlex.fi/en/laki/kaannokset/2004/en20041299.pdf> (date accessed 31.07.2014).
Act on Environmental Protection in Maritime Transport, (1672/2009), available at; <http://www.finlex.fi/en/laki/kaannokset/2009/en20091672.pdf> (date accessed 31.07.2014).
Environmental Protection act, (86/2000; amendments up to 728/2011 included), February 4, 2000, available at; <http://www.finlex.fi/en/laki/kaannokset/2000/en20000086.pdf> (date accessed 31.07.2014).
Environmental Protection Decree, No. 169/2000, Issued in Helsinki 18 February 2000, available at; <http://www.finlex.fi/en/laki/kaannokset/1999/en20000169.pdf> (date accessed 31.07.2014).
Finland's Strategy for the Arctic Region, Prime Ministers Office Publications, 5 July 2010, available at; <http://web.law.columbia.edu/sites/default/files/microsites/climate-change/files/Resources/FinlandArctic.pdf> (date accessed 30.09.2014).
Government Decree on Water Resources Management (1040/2006) Given in Helsinki on 30 November, 2006, available at <http://www.finlex.fi/en/laki/kaannokset/2006/en20061040.pdf> (date accessed 31.07.2014).
Government Decree on Water Resources Management Regions (1303/2004), Issued in Helsinki 30 December 2004, available at; <http://www.finlex.fi/en/laki/kaannokset/2004/en20041303.pdf> (date accessed 31.07.2014).
Government Decree on Substances Dangerous and Harmful to the Aquatic Environment 1022/2006, Issued in Helsinki 23 November 2006, available at; <http://www.finlex.fi/en/laki/kaannokset/2006/en20061022.pdf> (date accessed 31.07.2014).
Water Services Act, (119/2001), available at <http://www.finlex.fi/en/laki/kaannokset/2001/en20010119.pdf> (date accessed 31.07.2014).

Iceland

A Parliamentary Resolution on Iceland's Arctic Policy, Approved by Althingi at the 139th legislative session, March, 28 2011, available at; <https://web.law.columbia.edu/sites/default/files/microsites/climate-change/files/Arctic-Resources/Arctic-Council/04_Iceland%20Arctic-Policy-Approved-by-Althingi.pdf> (date accessed 29.07.2014).
The 2004 Act on Marine and Coastal Anti-Pollution, Act no. 33/2004 on marine and coastal anti-pollution measures.
The Nature Conservation Act no. 44/1999, Ministry for the Encironment of Iceland, available at <http://eng.umhverfisraduneyti.is/legislation/nr/389> (date accessed 31.07.2014).

Sweden

Act Relative to measures against pollution caused by ships (Translated version) [Förordning (1980:789) om åtgärder mot förorening från fartyg], available at; http://www.notisum.se/rnp/sls/lag/19800789.htm (date accessed 31.07.2014).

Sweden's Strategy for the Arctic Region, Regeringskansliet, Government offices of Sweden, 2011, available at; <https://web.law.columbia.edu/sites/default/files/microsites/climate-change/files/Arctic-Resources/Arctic-Council/Sweden%20Arctic%20Policy.pdf >, (date accessed 31.07.2014).

Norway

Act of 15 June 2001 No. 79 Relating to the Protection of the Environment in Svalbard, Amended by the Act of 20 April 2012 No. 20 (in force from 1 July 2012 in accordance with the Royal Decree of 20 April 2012 No. 332), available at; <http://www.regjeringen.no/en/doc/laws/acts/svalbard-environmental-protection-act.html?id=173945> (date accessed 31.07.2014).

The Norwegian Government's Hight North Strategy, the Norwegian Ministry of Foreign Affairs, 2006, available at; <http://www.regjeringen.no/upload/Ud/Vedlegg/strategien.pdf> (date accessed 31.07.2014).

European Union

Directive 2005/35/EC of the European Parliament and of the Council of 7 September 2005 on ship-source pollution and on the introduction of penalties for infringements. Procedure number: 2993/00377COD. European Parliament. Retrieved 30.04.2014 from http://eur-lex.europa.eu/.

Articles in Journals/Yearbooks

Abbott, Kenneth and Duncan Snidal, "Hard and Soft Law in International Governance," *International Organization*, 2000, Volume 54.

Baker, Betsy, "Filling an Arctic Gap: Legal and Regulatory Possibilities for Canadian– American Cooperation in the Beaufort Sea," *Vermont Law Review*, 2009 (34).

Boyle, Alan E., "Marine Pollution under the Law of the Sea Convention", *The American Journal of International Law*, 1985, Vol. 79.

Carroll, A.B. (1991) "The pyramid of corporate social responsibility: toward the moral management of organizational stakeholders", *Business Horizons*, Vol. 34, Iss. No. 4, pp. 39–48.

Chaturvedi, Chaturvedi, "Ocean Governance and the Polar Regions: Geopolitics, Law, and Sustainability", *15 Ocean Y.B. 475*, 2001, Centre for the Study of Geopolitics, Panjab University.

Corell, Hans, "Reflections on the Possibilities and Limitations of a Binding Legal Regime", *Environmental Policy and Law*, 2007, Vol. 37, No. 4.

Crowder, L.B. et al., "Resolving Mismatches in U.S. Ocean Governance," *Science*, 4 August 2006 (313), Vol. 313 no. 5787 pp. 617-618, DOI: 10.1126/science.1129706

Davis, K., "The Case for and against business assumption of social responsibilities", *Academy of Management Journal*, 1973, Volume 16, Issue No. 2.

De la Fayette, Louise, "The Marine Environment Protection Committee: The Conjunction of the Law of the Sea and International Environmental Law", *The International Journal of Maritime and Coastal Law*, 2001, Vol. 16, No 2.

Davis, K., "Five Propositions for Social Responsibility", *Business Horizon*, 1975, Volume 18, No. 3.

Douglas R. and Willy Østreng, Willy, "The Northern sea route regime: Exquisite superpower subterfuge?", *Ocean Development and International Law*, 1999, Vol. 30, Issue 4.

Doelle, Meinhard, "Climate Change and the Use of the Dispute Settlement Regime of the Law of the Sea Convention", *Ocean Development International Law*, 2006, Vol. 37, Issue: 3–4, Taylor & Francis Publishers.

Ebinger, Charles K. and Evie Zambetakis, "The geopolicis of Arctic melt," 2009, *International Affairs*, 85.

Emiliani, M. L., "A mathematical logic approach to the shareholder vs. stakeholder debate", *Management Decision*, 2001, Volume 39, No. 8.

Falkingham, John, Melling, Dr. Humfrey and Wilson, Katherine J., "Shipping in the Canadian Arctic: possible climate change scenarios", *Newsletter of the northern climate change*, 2002, p. 4, available at; <http://www.taiga.net/nce/resources/newsletters/NCE_Newsletter_Fall2002.pdf> (date accessed 13 March 2010).

Ford, James D., Smit, Barry and Wandel, Johanna, "Vulnerability to Climate Change in the Arctic: A case study from Arctic Bay, Canada" in *Global Environmental Change*, 2006, Vol. 6, Issue: 2, Elsevier Publisher.

Franckx, Erik, "The Legal Regime of Navigation in the Russian Arctic", *Journal of Transnational Law and Policy*, 2009, Vol. 18, no. 2.

Frederick, W. C., "Toward CSR sub 1 to sub 2: the maturing of business and society thought", *Business and Society*, 1978/1994, Volume 33, Issue No. 2.

Frederick, W. C., "Towards CSR sub 3. Why ethical analysis is indispensible and unavoidable in corporate affairs", *California Management Review*, 1986, volume 28, Issue No. 2.

French, Duncan, "Sustainable Development and the 1991 Madrid Protocol to the 1959 Antarctic Treaty: The Primacy of Protection in a Particularly Sensitive Environment", (1999), vol. 2:3, *JIWLP*.

Goodpaster, K. E., "Business ethics and stakeholder analysis", *Business ethics Quarterly*, 1991, Volume 1, issue No. 1.

Green, L. C., *The Modern Law Review*, Vol. 15, No. 3 (Jul., 1952).

Griffin, Andrew, "MARPOL 73/78 and Vessel Pollution: A Glass Half Full or Half Empty?", *Indiana Journal of Global Legal Studies*, 4 January 1994, Vol. 1, Issue 2, Article 10.

Halal, W.E., "Corporate community: a theory of the firm uniting profitability and responsibility", *Strategy and Leadership*, 2000, Vol. 28, Iss. No. 2.

Harrison, J. S. and Freeman, R. E., "Stakeholder social responsibility, and performance: empirical evidence and theoretical perspectives", *Academy and Management Journal,* 1999, Volume 44, No. 2.

Hess, D., Rogovsky, N. And Dunfee, T. W., "The next wave of corporate community involvement: corporate social initiatives", *Corporate Management Review*, 2002, Volume 44, No. 2.

Hillman, AJ. J., Keim, G. D., and Luce, R. A., "Board composition and stakeholder performance: do stakeholder directors make a difference?", *Business and Society*, 2001, Volume. 40, Issue. No. 3.

Huebert, Rob, "The Need for an Arctic Treaty: Growing from the United Nations Convention on the Law of the Sea", 2009, *Oceans Yearbook.*

Hummels, H., "Organizing ethics: a stakeholder debate", *Journal of Business Ethics*, 1998, Vol. 17, Issue No. 13.

Hummel, M., *Protecting Canada's Endangered Spaces: An Owner's Manual*, 1995, Key Porter Books Ltd., Toronto.

Johnston, D.M., "The Future of the Arctic Ocean: Competing Domains of International Public Policy", 17 *Ocean Yearbook*, 2003.

Jones, T.M., "Corporate social responsibility revisited, redefined", *California Management Review*, 1980, Vol. 22, No. 3.

Jones, T.M., "Instrumental stakeholder theory: a synthesis of ethics", *The Academy of Management Review*, 1999, Vol. 20, Iss. No. 2.

Jones, T. and Wicks, A., "Convergent stakeholder theory", *Academy of Management Review*, 1999, Vol. 24.

Juda, Lawrence, "Considerations in Developing a Functional approach to the Governance of Large Marine Ecosystems", *Ocean Development and International Law 30*, 1999.

Kakabadse, Nada K., Rozuel, C and Lee-Davies, L., "Corporate Social Responsibility and Stakeholder Approach: A Conceptual Review", *International Journal Business governance and Ethics*, Vol. 1, No. 4, 2005.

Kilcullen, M. And Ohles Kooistra, J., "At least do no harm: sources on the changing role of business ethics and corporate social responsibility", *Reference Service Review*, 1999, Volume 27, No. 2.

Kaldor, Mary, "The Idea of Global Civil Society", *International Affairs*, 2003, Volume 70.

Koivurova, Timo, "Alternatives for an Arctic Treaty-Evaluation and a New Proposal", *Review of European Community and International Environmental Law* RECIEL, 17 (1) 2008, Special International Polar Year Issue.

Kraska, James, "The Law of the Sea Convention and the NorthWest Passage" in *International Journal of Marine and Coastal Law*, 2007, Vol. 22, No. 2, 9 May.

Kullenberg, G., "Approaches to addressing the problems of pollution of the marine environment", *Ocean and Coastal Management*, 1999, Volume 42, Number 12, December.

Lamson, Cynthia, "Arctic Shipping, Marine Safety and Environmental Protection", *Marine Policy*, 1987, Vol. 11.

Lantos, G.P., "The boundaries of strategic corporate social responsibility", *Journal of Consumer Marketing*, 2001, Volume 18, No. 7.

Levy, March A.,"Is the Environment a National Security Issue?", *International Security*, Vol. 20, No. 2, Autumn 1995, MIT Press.

Lipson, Charles, "Why Are Some International Agreements Informal", *International Organization*, 1991, Volume 45, Issue 4.

L'Etang, J., "Ethical corporate Social Responsibility: a Framework for Managers", *Journal of Business Ethics*, 1995, Volume 18, no. 7.

Marffy, Annick de, "Ocean Governance: A Process in the Right Direction for the Effective Management of the Oceans", *18 Ocean Y.B. 162*, 2004, Division for Ocean Affairs and the Law of the Sea, Office of Legal Affairs, United Nations, New York.

Margolis, J.D. and Walsh, J. P., "Misery loves companies: rethinking social initiatives by business", *Administrative Science Quarterly*, 2003, Volume 48.

McAdam, R. and Leonard, D., "Corporate social responsibility ina total quality management context: opportunities for sustainable growth", *Corporate Governance*, 2003, Volume 3, No. 4.

McDorman, Ted L., "In the Wake of the 'Polar Sea': Canadian Jurisdiction and the Northwest Passage", *Les Cahiers de Droit*, 1986, Vol. 27.

McKinnon, J.B., "Arctic Baselines: *A Litore Usque ad Litus*", in 66 *Can. Bar Rev*, 1987.

McRae, D.M. and Goundrey, D.J., "Environmental Jurisdiction in arctic waters: the extent of article 234", *University of British Columbia Law Review*, 1982, Vol. 16:2.

McWilliams, A. and Siegel, D., "Corporate social responsibility: a theory of the firm perspective", *The Academy of Management Review*, 2001, Volume 26, Issue No. 1.

Mitnick, B.M., "Systematics and CSR: the theory and processes of normatice referencing", *Business and Society*, 1995, Volume. 34, Issue No. 1.

Molenaar, E.J., Book Review of "Coastal State Regulation of International Shipping" by L.S. Johnson, 22 *International Journal of Marine and Coastal Law*, 2007.

Moir, L., "What do we mean by corporate social responsibility?", *Corporate Governance Ethics*, Volume 16, Issue No. 4.

Morris, S. A., "Systematic and CSR: the theory and processes of normative referencing", *Journal of Business Ethics*, 1997, Volume 16, issue. No. 4.

Mitnick, B.M., "Systematics and CSR: the theory and processes of normative referencing", *Business and Society*, 1995. Vol. 34, Iss. No. 1.

Oldfield, J.D., "Russian Environmentalism", *European Environment: the Journal of European Environmental Policy*, 14 March 2002, doi: 10.1002/eet.286, Vol. 12, No. 2.

Petersen, Nikolaj, "The Arctic as a New Arena for Danish Foreign Policy: The Ilulissat Initiative and its Implications," *Danish Foreign Policy Yearbook* 2009. Copenhagen.

Post, J.E., Preston, L.E. and Sachs, S., "Managing the extended enterprise: the new stakeholder view", *California Management Review*, 2002, Vol. 45, No. 1.

Punter, L. and Gangneux, D., "Social accountability: the most recent element to ensure total quality management", *Journal of Total Quality Management*, 1998, Volume 9, no. 4.

Pharand, Donat, "The legal regime of the Arctic: Some outstanding issues", *International Journal*, 1984, Vol. 34, No. 9.

Pharand, Donat, "The Arctic Waters and the Northwest Passage: A Final Revisit", 38 *Ocean Development & International Law,* 2007, Accessed via Elin@Lund (date accessed 13 March 2012).

Rayfuse, Rosemary, "Melting Moments: The Future of Polar Oceans Governance in a Warming World", *Review of European Comminity and International Environmental* Law, 2007 (16).

Robertson, D. C. and Nicholson, N., "Expressions of corporate social responsibility in U.K. firms", *Journal of Business Ethics*, 1996, Volume 15, Issue No. 10.

Roginko, Alexei Yu and LaMourie, Matthew J., "Emerging marine environmental protection strategies for the Arctic", *Marine Policy*, 1992, Vol. 16, Issue 4.

Rothwell, Donald R., "The Canadian-U.S. Northwest Passage Dispute: A Reassessment", 1993, 26 *Cornell Int'l L.J.*

Rubenstein, Daniel B., "Collaborative Oceans Governance: Measuring Efficacy", *20 Ocean Year Book 147*, 2006.

Scrivener, David, "Gorbachev's Murmansk Speech: Soviet Initiatives and Western Responses," *Security Policy Library No.* 1, 1989, Oslo.

Scrivener, David, "Environmental Cooperation in the Arctic: From Strategy to Council", *Security Policy Library No.1*, (Oslo: Norwegian Atlantic Committee, 1996).

Scott, D. and Suffling, R., *Climate Change and Canada's, National Park System: A Screening Level Assessment*, Environment Canada and Parks Canada, 2000, Hull, Quebec.

Sethi, S.P., "Dimensions of corporate social performance: an analytic framework", *California Management Review*, 1975, Volume 17, No. 3.

Forster, Malcolm J., "The Prevention of Oil Pollution Act 1971", *The International and Comparative Law Quarterly*, 1972, Vol. 21, No. 4.

Simmons, J., "Managing in the post-managerialist era: towards socially responsible corporate governance", 2004, *Management Decision*, Vol. 42, Nos. 3–4.

Singhapakdi, A., Vitell, S. J., Rallapalli, K.C. and Kraft K. L., "The perceived role of ethics and social responsibility: a scale development", *Journal of Business Ethics,* 1996, Volume 15, Issue No. 11.

Spinnato, J.M., "Historic and Vital Bays: An Analysis of Libya's Claim to the Gulf of Sidra", 13 *Ocean Development and International Law Journal*, 1983.

Stokke, Olav Schram, "A legal regime for the Arctic? Interplay with the Law of the Sea Convention", *Marine Policy,* 2007, Vol. 31.

Sutcliffe, W.H. Jr., Loucks, R.H., Drinkwater, K.F. and Coote, A.R., "Nutrient flux onto the Labrador Shelf from Hudson Strait and its biological consequences", *Canadian Journal of Fisheries and Aquatic Sciences*, 1983, 40,

Szwajkowski, E., "Simplifying the principles of stakeholder management: the three most important principles", *Business and Society*, 2000, Vol. 39, Iss. No. 4.

Takala, T., "Ownership, responsibility and leadership—a historical perspective", *International Journal of Social Economics*, 1999, Volume 26, No. 6.

Thevenet, M., "Global Responsibility and Individual Exemplarity", *Corporate Governance*, Volume 3, No. 3.

Thomas, A. S. and Simerly R. I., "Internal determinants of corporate social performance: the role of top managers", *Academy of Management Journal*, 1995, Special Issue –best paper proceedings.

Utton, Albert E., "Protective Measures and the Torrey Canyon", *9 B.C.L. Rev. 613*, 1968, available at <http://lawdigitalcommons.bc.edu/bclr/vol9/iss3/4>, (date accessed 27 February 2012).

VanderZwaag, David, Huebert, R. and Ferrara, Stacey, "The Arctic Environmental Protection Strategy, Arctic Council and Multilateral Environmental Initiatives: Tinkering while the Arctic Marine Environment Totters", *Denver Journal of International Law and Policy*, Vol. 30:2.

Vinten, G., "The Stakeholder Manager", *Management Decision*, 2000, Volume. 38, Issue: 6.
West, Mary B., "Arctic Warming: Environmental, Environmental and Security Implications", *Vanderbilt Journal of Transitional Law*, 2012, Vol. 42:1081.
Wheeler, D and Sillanpää, M., "Including the Stakeholders: the business case", *Long range Planning*, 1998, Volume 31, No. 2.
Wilson, I., "The New Rules: ethics, social responsibility and strategy", *Strategy and Leadership*, 2000. Vol. 20, Issue: 3.
Wood, D.J., "Corporate social performance revisited", *The Academy of Management Review*, 1991, Vol. 16, No. 4, pp. 691–718.
Yankov, Alexander, "The Concept of Protection and Sustainable Development of the Marine Environment", *18 Ocean Y.B. 267*, 2004.
Young, Oran R., "Institutional Linkages in International Society: Polar Perspectives," 1996, *Global Governance* (2).
Young, Oran R., "Arctic Governance—Pathways to the future", *Arctic Review on Law and Politics*, vol. 1, 2/2010 pp. 164–185. ISSN 1891-6252.
Zysk, Katarzyna, "Russia's Arctic Strategy: Ambitions and Constraints", *Joint Force Quarterly*, Issue 57, 2nd Quarter, 2010, available at: <http://www.ndu.edu/press/lib/images/jfq-57/zysk.pdf> (date accessed 20 February 2012).

Magazines, Newspapers and Webpages

CBC News, *Non-northern nations get more interested in Arctic Council*, available at; <http://www.cbc.ca/canada/north/story/2008/04/11/arctic-council.html> (date accessed 31.07.2014).
Funk, Mckenzie, Arctic Landgrab, *National Geographic*, Vol. 215, No. 5, May 2005.
Graff, James, Fight for the top of the World, *Time Magazine*, Vol. 170, No. 13, October 2007.
Greenpeace International website, available at; <www.greenpeace.org>.
Norden, About Nordic Co-operation, available at; < http://www.norden.org/en>, (date accessed 27 July 2014).
Official Webpage of IMO, available at; <http://www.imo.org> (date accessed 23 February 2012).
Official Webpage of OSPAR, available at; <http://www.ospar.org> (date accessed 27 February 2012).
Official Website of the European Union Parliament, available at; <http://www.europarl.europa.eu/sides/getDoc.do?type=TA&reference=P6-TA-2008-0474&language=EN>, (date accessed 27 July 2014).
Official website of U.S. Geological Survey, *90 Billion Barrels of Oil and 1,670 Trillion Cubic Feet of Natural Gas Assessed in the Arctic*, available at; <http://www.usgs.gov/newsroom/article.asp?ID=1980&from=rss_home#.U9SsiIB_u24> (date accessed 27 July 2014).
Official Website of the Danish Maritime Authoirty, available at; <http://www.dma.dk/Policy/Sider/Mainpage.aspx>, (date accessed 29.07.2014).
Russian Federation Government, "Basics of the State Policy of the Russian Federation in the Arctic for the Period Till 2020 and for a Further Perspective," publication of the official governmental newspaper *Rossiyskaya Gazeta*, 30 March 2009.
The Times (London), March 29, 1967, at p. 1, column 1.

Other Documents (Reports, Bulletins, Commentaries and Proceedings of Conferences, Bi-lateral Agreements, Documents etc.)

Agreement between the government of Canada and the government of the U.S. on Arctic cooperation (Canada-U.S.), 11 January 1988, C. T. S. 1988/29, available at; <http://www.lexum.umontreal.ca/ca_us/en/cts.1988.29.en.html> (date accessed 20 March 2012).

Arctic TRANSFORM, a project of Ecologic, Background Papers, 2008, Berlin.

Bill C-3: An Act to Amend the Arctic Waters Pollution Prevention Act, Legislative Summary, LS-617E, Penny Becklumb Industry, Infrastructure and Resources Division, 19 December 2008, Accessed at <http://www.parl.gc.ca/Content/LOP/LegislativeSummaries/40/1/401c3-e.pdf> (date accessed 22 February 2012).

Brigham L. and McCalla R., *Arctic Marine Shipping, 2009 Report.* Arctic Marine Shipping Assessment (AMSA).

Canadian Coast Guard, *The Northwest Passage: Summary of Complete Transits, Fleet News no. 11*, 1985.

Chapin, F.S., M. Hoel, S.R. Carpenter, J. Lubchenco, B. Walker, et al., "Building Resilience and Adaptation to Manage Arctic Change," *Ambio*, 2006 (35).

Conley, Heather and Kraut, Jamey, *U.S. Strategic Interests in the Arctic: An Assessment of Current Challenges and New Opportunities for Co-operation*, "A Report of the CSIS Europe Program", Centre for Strategic & International Studies, April 2010.

Csonka, Yvon & Schweitzer, Peter, *Societies and Cultures. Change and Persistence*, in Arctic Human Development Report 45, 47, 53, 2004, Niels Einarsson et al. eds.

Elliot, Michael, "Biological pollutants and biological pollution-an increasing cause for concern", *Marine Pollution Bulletin*, 2003, volume 46:1, Publications expediting incorporation.

Environmental Response, Report of the Canadian Coast Guard Maritime Services, Ottawa, Ontario, 2009, Fisheries and Oceans Canada Publishers, available at; <www.ccg-gcc.gc.ca/eng/Ccg/er_Home> (date accessed 2 March 2012).

Europa press release available at; <http://europa.eu/rapid/pressReleasesAction.do?reference=MEMO/05/437&format=HTML&aged=1&language=EN&guiLanguage=en> (date accessed 29 February 2012).

Financing Sustainable Development: GEF and the "Northern Dimension" Remarks by Mohamed T. El-Ashry Chief Executive Officer and Chairman Global Environment Facility, Rovaniemi, Finland, August 29, 2000.

Germanischar Lloyd: Annual Report 2006, *Oil and Ice: Market leader for ships with ice class/Standards developed for navigating in ice*, 2006.

"Governance of Arctic Shipping" in *Arctic Marine Shipping Assessment*, 29 April 2009, available at <http://www.arctic.gov/publications/AMSA/governance.pdf> (date accessed 23 February 2012).

Hubert R. and B. Yeager, Brooks B., *A New Sea: The Need for a Regional Agreement on Management and Conservation of the Arctic Marine Environment*, January 2008, A Report Published by WWF International Arctic Programme, Oslo, Norway, available at; <http://www.file:///Users/tafsir/Downloads/WWF%252BA%252BNew%252BSea%252B-%252B%252BReport%252BOctober%252B2006.pdf> (date accessed 6 August 2014).

Interagency Ocean Policy Task Force, "Interim Framework for Effective Coastal and Marine Spatial Planning", 9 December 2009, available at <http://www.whitehouse.gov/sites/default/files/microsites/091209-Interim-CMSP-Framework-Task-Force.pdf> (date accessed 6 August 2014).

Inuit Circumpolar Conference, Towards an Inuit Regional Conservation Strategy, 1996; Agenda 21 from an Inuit Perspective.

Jensen, Øystein, *The IMO Guidelines for Ships Operating in Arctic Ice-Covered Waters: From Voluntary to Mandatory Tool for Navigation Safety and Environmental Protection?* The Fritjof Nansen Institute, FNI-rapport 2/2007.

Ker, Alex, The Legal Regulatory and Policy Framework for Non-renewable Resource Development in the Northwest Territories, (Ottawa: National Round Table on the Environment and the Economy), 2000.

Kitagawa, Hiromitsu, "Arctic sRouting: Challenges and Opportunities" in *Impacts of Climate Change on the Maritime Industry*, The proceedings of the Conference on Impacts of Climate Change on the Maritime Industry, 2–4 June 2008 (Sweden), Neil Bellefontaine and Olof Linden (eds.), 2009, World Maritime University Publications.

Koivurova, Timo and Erik J. Molenaar, "International Governance and Regulation of the Marine Arctic: A Proposal for a Legally Binding Regime", WWF International Arctic Programme, 2010, Oslo.

Legal Aspect of Arctic Shipping, *Legal and socio-economic studies in the field of the Integrated Maritime Policy for the European Union*, Summary Report of the European Commission, 2010, European Union. ISBN 978-92-79-15828-5, doi: 10.2771/51132.

Mulherin, Nathan D., *The Northern Sea Route: its development and evolving state of operations in the 1990s / Nathan D. Mulherin ; prepared for U.S. Army Engineer District, Alaska*, 1996, U.S. Army Corps of Engineers, Cold Regions Research & Engineering Laboratory.

Nowlan, Linda, *Arctic Legal Regime for Environmental Protection*, 2001, IUCN, Gland, Switzerland and Cambridge, UK and ICEL, Bonn, Germany, available at; <http://weavingaweb.org/pdfdocuments/EPLP44EN.pdf> (date accessed 31 March 2012).

Pagnan, Jeanne L. *Arctic Marine Protection*. InfoNorth, vol. 53, No. 4, December 2000, available at; <http://www.arctic.ucalgary.ca/index.php?page=arctic_contents> (date accessed 31.07.2014).

PAME: Working Group on the Protection of the Arctic Marine Environment, Report to the Third Ministerial Conference on the Protection of the Arctic Environment, 20-21 March 1996 Inuvik, Canada, Ministry of Environment.

Perry, Robin Robin Perry, *Protection of Sensitive Coastal Areas in the United Kingdom: FromTorrey Canyon to the New Millenium*, International Oil Spill Conference Proceedings: March 1999, Vol. 1999, No. 1, pp. 149–163.

Quality Status Report 2000 for the North-East Atlantic, available at; <http://www.ospar.org/eng/html/welcome.html> (date accessed 31.07.2014).

Report of the United Nations Secretary General, "Impact of the entry into force of the 1982 United Nations Convention on the Law of the Sea on related, existing, and proposed instruments and programmes", UN Doc. A/52/491, Section J.

Borgerson, Scott G., "Arctic Meltdown: The Economic and Security Implications of Global Warming", 2008, *Foreign Affairs*, Published by the Council on Foreign Relations, available at; <http://www.foreignaffairs.com/articles/63222/scott-g-borgerson/arctic-meltdown>

State of the Arctic Environment Report (SOAER), 1997, Arctic Council.

Testimony of Oran Young to SCFAIT, Canada and the Circumpolar World, Chap. 5.

The text of the Antarctic Treaty is published in the United Nations Treaty Series, vol. 402, 71, available at; <https://treaties.un.org/Pages/FileSearch.aspx?tab=SEARCH> (date accessed 6 August 2014).

Transport Canada Seaway and Domestic Shipping Policy, 2005, Canadian Arctic Shipping Assessment Scoping Study.

United Nations Conference on Environment & Development, Rio de Janerio, Brazil, 3 to 14 June 1992, Agenda 21, United Nations Sustainable Development.

United States Arctic Research Commission [USARC], 2006, Annual Report of the United States Arctic Research Commission to the President and Congress of the United States: Fiscal year 2006.

UNEP Organization Profile, available at; <http://www.unep.org/PDF/UNEPOrganizationProfile.pdf> (date accessed 31.07.2014).

UNEP, GEO-2000 Global Environment Outlook, (UNEP: Nairobi), 2000, Chap 2, The Antarctic.

UNEP, *Global Programme of Action for the Protection of the Marine Environment from Land-based Activities*. Adopted on 3 November 1995, available at; <http://www.gpa.unep.org/documents/full_text_of_the_english.pdf> (date accessed 31.07.2014).

WWF. *A New Sea. The need for a regional agreement on management and conservation of the arctic marine environment*. Published by WWF International Arctic Programme, Oslo, 2008, available at; <http://assets.panda.org/downloads/a_new_sea_jan08_final_11jan08.pdf> (date accessed 31.07.2014).

WWF, Indigenous and Traditional Peoples of the World and Ecoregion Conservation: An Integrated Approach to Conserving the World's Biological and Cultural Diversity, 2000.

World Commission of Environment and Development, *Our Common Future*, 1987, New York: Oxford University Press.

WWF, *A New Sea: The need for a regional agreement on management and conservation of the arctic marine environment*. Published by WWF International Arctic Programme, Oslo, 2008, available at; <www.vliz.be/imisdocs/publications/130977.pdf> (date accessed 13 April 2012).

Young, Oran R., *The Structure of Arctic Cooperation: Solving Problems/Seizing Opportunities*, Fourth conference of Parliamentarians of the Arctic Region, Rovaniemi, 2000, available at; <http://www.arcticparl.org/_res/site/File/images/conf4_sac.pdf> (date accessed 30 March 2012).

1996 Utrecht ATCM Report, Final Report of the XX Antarctic Treaty Consultative Meeting, available at; <http://www.anta.canterbury.ac.nz/resources/treaty/meetings/xx-utrecht/p1final.html> (date accessed 6 August 2014).

1996 Report of the Auditor General of Canada, Chap. 22—Main Points–Federal Contaminated Sites—Management Information on Environmental Costs and Liabilities; 1999 Report of the Commissioner of the Environment and Sustainable Development, (Government of Canada: Ottawa).

1997 Christchurch ATCM Report, Final Report of the XX I Antarctic Treaty Consultative Meeting, available at; <http://www.ats.aq/devAS/ats_meetings.aspx?lang=e> (date accessed 6 August 2014).

MIX
Papier aus verantwortungsvollen Quellen
Paper from responsible sources
FSC® C105338

FSC
www.fsc.org

If you have any concerns about our products,
you can contact us on
ProductSafety@springernature.com

In case Publisher is established outside the EU,
the EU authorized representative is:
Springer Nature Customer Service Center GmbH
Europaplatz 3, 69115 Heidelberg, Germany

Printed by Libri Plureos GmbH
in Hamburg, Germany